THE BIO-ORIGINS SERIES

# Origins of Sex

*Three Billion Years of Genetic Recombination*

LYNN MARGULIS

DORION SAGAN

YALE UNIVERSITY PRESS
New Haven and London

Designed by Margaret E.B. Joyner
and set in Caledonia type by The Composing Room of Michigan.
Printed in the United States of America by Halliday Lithograph,
West Hanover, Massachusetts.


**Library of Congress Cataloging in Publication Data**

Margulis, Lynn, 1938–
  Origins of sex.

  (The Bio-origins series)
  Bibliography: p.
  Includes index.
  1. Sex (Biology)   2. Genetic recombination.   I. Sagan,
Dorion, 1959–  .   II. Title.   III. Series.
QH481.M27   1986      575.1      85-8385
ISBN 0–300–03340–0 (alk. paper)


The paper in this book meets the guidelines for
permanence and durability of the Committee on
Production Guidelines for Book Longevity of the
Council on Library Resources.


10   9   8   7   6   5   4   3   2   1

For Tonio and his generation;

for the combinations,

sexual and parasexual,

that bring us out of ourselves

and make us more than

we are alone.

# CONTENTS

# FIGURES

# TABLES

# ACKNOWLEDGMENTS

We acknowledge with gratitude useful discussions on aspects of the origins of sex with our colleagues: David Bermudes, Betsey Dyer, Richard Dawkins, Michael Dolan, Pamela Hall, Gregory Hinkle, Donna Mehos, Heather McKhann, Robert Obar, Lorraine Olendzenski, Mark Ridley, and John Stolz. We thank John Langridge for his useful critical review of the manuscript and Harlyn Halvorson for arranging the Origin of Sex symposium at the Marine Biological Laboratory (August 1984). We are grateful to Lorraine Olendzenski for representing us at that meeting and for her extensive work in preparing the manuscript. Although the drawings by Laszlo Meszoly, Christie Lyons, and Steven Alexander speak for themselves, we are delighted with our continued opportunity to work with these artists.

We are in the debt of two scholars: the prescient L. R. Cleveland (1892–1969), who attempted, without success, to make views very much like these known; and Igor Raikov, working today at the Institute of Cytology, Leningrad, who for many years has painstakingly collected information relevant to the sex life of protists.

Finally, we are overwhelmingly grateful to our perceptive and appropriately critical editor, Edward Tripp, who read the manuscript twice, each time adding his wisdom to it, and with whom at all stages of preparation it was a pleasure to work. We are also very grateful to Kate Schmit and Barbara Hofmaier for their excellent editorial assistance.

Aspects of the research were supported by the NASA Life Science Office (under grant NGR-004-025 to L. M.) and by the Boston University Graduate School.

# INTRODUCTION

## For Whom This Book Is Meant

This book is meant to be an evolutionary detective story that unravels the mystery and history of the origin of sex. Different forms of sex are observed in the biological world. We want to know how sex came to be.

Sex in bacteria crosses species boundaries and allows a flow of genetic information that some consider the basis for a worldwide gene pool composed of bacteria (Sonea and Panisset, 1983). Other organisms, those with nuclei, probably evolved through endosymbiosis: bacteria living inside each other shared each other's foods, metabolites, and eventually genes (Margulis, 1981). Nonbacterial, meiotic sex, found only in nucleated organisms, has a different origin and history. Although less important as a raw source of variety for natural selection, it is crucial to the development and reproduction of animals and plants. Our hypothesis is that the history of the meiotic sex of animals and plants depended on cannibalism in protists and the differential replication of their former symbionts, some of which became organelles of these protists. The existence of modern-day protists, which show far more variation on the theme of meiotic sex than any other living group, suggests that sex evolved in their ancestors. Studying sexuality in extant organisms has led us to conclude that meiotic sex became coupled to reproduction in animals and plants only because differentiation of their cells was impossible without it.

To reconstruct the events leading to the origin of sexuality is a difficult task, because the essential cellular events at its basis are so ancient and because they occurred in microorganisms that did not preserve well in the fossil record. In evolution, as in criminology, one is never absolutely sure about a given reconstruction. Nonetheless, it is our pleasure here to provide a scenario for the origins of sex that we feel is consistent with the mass of circumstantial evidence so far accumulated.

1

Everyone is interested in sex. But, from a scientific perspective, the word is all too often associated with reproduction, with sexual intercourse leading to childbirth. As we look over the evolutionary history of life, however, we see that sex is the formation of a genetically new individual. Sex is a genetic mixing process that has nothing necessarily to do with reproduction as we know it in mammals. Throughout evolutionary history a great many organisms offered and exchanged genes sexually without the sex ever leading to the cell or organism copying known as reproduction. Although additional living beings are often reproduced by a contribution of genes from more than a single parent, sex in most organisms is still divorced from growth and reproduction, which are accomplished by nonsexual means.

Biologically, sex is part of the rich repertoire of life. Any specific instance of a sexual event is complex. Each event in a sexual process—for example, fertilization in a plant or animal—has its own specific history. Originally unrelated phenomena, such as genetic exchange (as in DNA recombination) and cell reproduction, often became entangled after having evolved from separate beginnings. The story of sex starts with an account of the earliest life on Earth. The private activities of early cells are involved even today in courtship among human beings. The intimate behavior of single cells has simply been elaborated to include animals and their behaviors and societies. Mammalian sex is a very late and special variation on a far more general theme.

The origin of sex is a problem that has long perplexed. It lends itself to innovative mythmaking (mythopoiesis); many cultures have imagined a primordial unisexual oneness that, under the influence of a celestial personality, was split into light and dark, heaven and earth, male and female, and so on. In the march of knowledge, however, mythical accounts of the origin of sex have been abandoned. We now realize, thanks to the insights of Darwinian evolution, that the sexual differences that loom so large in the daily lives of men and women did not arise at some specific time in the history of the human species. Evolution takes us far beyond the origin of apes and men, who at their first appearance were undoubtedly already fully sexual. Sex itself arose even earlier than the many species of sexual creatures with which we are familiar. It was present on the Earth when microbes, organisms that cannot be seen without a microscope, totally dominated the planetary surface. Sex was here for hundreds of millions of years before the first animals or plants appeared.

What keeps organisms that have sexual differences from devolving into the asexual state is, as we shall show, a completely different matter from how sex came about in the first place. Biologists, although they have tried, have not been able to prove that sexual organisms have an intrinsic advantage over asexual ones. Many have struggled with the question of how sexual organisms can afford to expend the biological "cost" of mating in every generation. Asexual organisms, since they can have more offspring per unit time, are, in Darwinian terms, more "fit." This sort of analysis implies that sexuality should disappear. But in animals sexuality is tenaciously maintained. We show here that this problem of the *maintenance* of sex (that which keeps animals and plants from becoming asexual) must be clearly distinguished from the problem of the *origins* of sex (the ways in which sex first evolved). There has been some confusion between these two aspects of sexual theory. The mix-up between remaining sexual and becoming sexual is one which we will try to steer well clear of throughout this book.

The origin of sex was not a one-time event. Sex is not a singular but a multifaceted and widespread phenomenon; it has developed several times, at the very least. The two most consequential appearances of sex were in tiny microbes—a half to about five micrometers long. Sex first appeared in bacteria. Later, in larger, more complex microbes called "protists," a new and different kind of sex evolved. Sex in bacteria is a biological mixing and matching on the molecular level: the splicing and mending of DNA molecules. Bacterial sexuality is very different from the meiotic sex of protists, fungi, plants, and animals, and it evolved far earlier. Meiosis, or cell division resulting in reduction in the number of chromosomes, and subsequent fertilization, or reunion of cells to reestablish the original chromosomal number, first occurred in protists. Protists, microbes generally from ten to a hundred micrometers long, are ancestral to fungi, animals, and plants. As protists evolved and gave rise to these other groups of organisms, sex was preserved. From a cellular vantage point, human sex is almost identical to that of some of the protistan microbes.

The main thesis of this book may come as a surprise to some. It is that ultimately males and females are different from each other not because sexual species are better equipped to handle the contingencies of a dynamically changing environment but because of a series of historical accidents that took place in and permitted the survival of ancestral protists. From the beginning the cellular events required for the emergence of differentiation—of cells, tissues, organs, and organ systems—were tied to meiotic sexuality. Conjugation and meiosis were intrinsic to the life cycles of differentiated protists, the ancestors of animals, fungi, and plants. We believe, from inferring the events necessary for the formation of the first sexual beings, that biparental sex itself did *not* immediately confer any great advantage upon those organisms in which it arose. This idea runs against the grain of the most popular rationale for the existence of sex: that sexual organisms, being on the average more diverse because they combine divergent traits, are more adaptable to changing environments. We do not think there is any evidence to justify the claim that sexual organisms are more diverse and therefore better equipped to cope with the vicissitudes of existence, nor that they reproduce in a sexual fashion because this permits them to "evolve faster." We think sexual organisms reproduce in this peculiar and "costly" fashion because, in the first place, they must reproduce and, second, because certain peculiarities of their evolutionary past have linked their reproduction and tissue differentiation with their sexuality. It is true that natural selection has favored many sexual organisms, but not *because* they were sexual (see chapter 11).

It is natural for human beings—who are mammals in which sex and reproduction are always associated—to think that the "purpose" of sex is reproduction. But in most microbes, organisms from which ultimately we have descended, sex is quite separate from reproduction. Reproduction is obligatory. All organisms reproduce; sex, on the other hand, is optional. Bacteria often engage in sex but they do not reproduce by cell division as a result of the sexual engagement. (Bacterial sex is not a precursor for the act of reproduction itself, but sometimes it supplies genes for survival at a given moment.) Many species of protists, fungi, animals, and plants can reproduce with or without two-parent sex. In some organisms sex, in fact, is lethal. It produces no offspring and destroys all those that participate in the process.

Thus the origin of sex and its subsequent maintenance is a complex subject whose history stretches back eons—to the trading of genes, in the form of DNA molecules in bacteria, and to the doubling up of chromosome sets in the amoebalike protists. The development of visible and consistent male/female differences, which occurred more than once and long after the origin of two-parent sex in identical-looking microbes, cannot be pinpointed at a definite time in the history of some prototypical ancestral plant or animal. Sexual origins involved many biochemical and genetic innovations, but they were first *microbiological* phenomena. We attempt to show here that sexual origins are encompassed not by one but by a series of happenings in symbiotic, sometimes starving microbes. All these historical happenings, related to the imperative of survival, took place in microbes more than a billion years ago. Sex has been preserved not because it is "adaptive" but because the organisms in which it was coupled to reproduction reproduced. Human lovers, male and female, are evolutionary permutations, living reminders of the ancient microbial events comprising the origin of sex. The pleasure we experience during orgasm, for instance, seems distinct and removed from the original activities that led to reproduction of single fused cells, but sexual climax is a feedback mechanism refined by natural selection: it induces us to carry on. The pursuit of sexual pleasure or love is, according to the biological point of view presented here, an elaborate device ensuring maintenance of the genetic system in the mammalian prelude to reproduction. Human sexual pleasure has more to do with reproduction than with sex. The events surrounding conception, during which single cells merge when a single egg is fertilized by a single sperm, are deeply entangled in the life history of our bodies. Today the repetition of these events is mandatory to the process of our reproduction.

Sex, in the form of nuclear fusion and chromosome reassortment in protists, is ancient. On the other hand, specific structures and behaviors associated with sex, such as large antlers in cervids and their use in spring rutting, are very recent appearances on the evolutionary scene. By attempting to place every structure, process, and behavior in an evolutionary context, we address the complex problem of the evolution of sex. When, in what population of organisms, under what selective pressures did a particular sex-related phenomenon evolve? Each case is different. We feel that an appreciation of evolutionary context is indispensable to understanding the origin and development of sex. Certain cellular aspects of sexuality are nearly identical. Nonetheless, many, indeed the vast majority of sexual manifestations that conspire to bring together mates are analogous rather than evolutionarily homologous. That is, similar sexual systems are not usually related through descent but evolved independently. This is why we must be explicit about the groups of organisms we are discussing when we refer to sexuality and its origins.

Although this book is intended as a straightforward narrative, its thesis represents a radically different "thought style" from that of the scientific literature from which it has emerged. A few words about the epistemological basis of scientific narrative are relevant here. The concept of thought styles in science has been developed by Ludwik Fleck (1979). Fleck claims, and we agree, that a thought style shared by members of a "thought collective" determines the formulation of every concept that underlies observation and description. "If we define the 'thought collective' as a community of persons mutually exchanging ideas or maintaining intellectual interaction, we will find by implication that it also provides the special 'carrier' for the historical development of any field of thought, as well as for the given stock of knowledge and level of culture. This we have designated thought style" (p. 39). Writing in 1935, Fleck recognizes that once "a structurally complete and closed system of opinions consisting of many details and relations has been formed, it offers enduring resistance to anything that contradicts it." He goes on to claim that words and ideas were not originally just *the names* for objects;

words were originally phonetic and mental equivalents of the experiences coinciding with them. In this way Fleck explains "the magical meaning of words" and the "dogmatic, reverential meaning of statements" (p. 27).

A thought collective consists of individuals, Fleck notes, but it is not simply the aggregate sum of them. Indeed, he asserts that one is hardly ever aware of the prevailing thought style in which one is operating. Although scientific thought styles should be more open than, say, religious ones, the dominant thought style—in this case the common view of the purpose and origin of sex—"almost always exerts an absolutely compulsive force upon [an individual's] thinking . . . with which it is not possible to be at variance" (p. 41). Using as his central example the spirochetal etiology of syphilis and its relation to the Wassermann diagnostic test, Fleck beautifully illustrates the social character of all scientific work. "This social character inherent in the very nature of scientific activity is not without its substantive consequences. Words which formerly were simple terms become slogans; sentences which once were simple statements become calls to battle" (p. 43).

The evolutionary origins of sex were far more complex and far less direct than the narrative we offer here. Indeed, our book is written in a different thought style from that reflected in preceding books and articles on this subject. Intrinsic to our viewpoint, as we explain in detail, are several concepts relatively new to the biological literature. For example, we take as given the concept that the largest discontinuity in life is that between the prokaryotic and the eukaryotic organisms. In addition, we assume that nucleated (eukaryotic) cells began as bacterial (prokaryotic) symbioses and thus today are legacies of coevolved microbial communities. We notice that individuals (with the exceptions of certain pure cultures of bacteria found only in the laboratory) are intrinsically composite structures and that sexuality and its origins have not been amenable to analysis until now because the paradoxical nature of "individuals" was not adequately known. We have consistently avoided the thought style of our colleagues, which has imposed a peculiar politicoeconomic terminology on the biology of sexuality. We perceive the prevalent analysis of sex in the biological literature (for instance, analyses in terms of "cost-benefit" or "parental investment") as a superb example of a Fleckian "thought style." Those who use this jargon are members of the thought collectives of sociobiology, population genetics, or population ecology, all of which are descendants of zoological tradition. We feel that zoological tradition is extraordinarily ill equipped for an analysis of the origins and maintenance of sex because most enlightening information comes from as yet poorly defined fields of science (thought collectives) outside zoology. For example, protistan nuclear cytology (Raikov, 1982), endocytobiology

(Schwemmler, 1980), and comparative protistology (Corliss, 1984) all provide starting points far more useful for the analysis of our problem than those offered by zoology and population biology. For these reasons we anticipate Fleckian "enduring resistance"—in other words, heavy attack of our book by the old guard. However, even if it excites verbal reprisal, our analysis ought to be of interest to all professionals concerned not only with the origin and maintenance of sexuality but also with the ways in which we must verify or reject our hypotheses on the basis of newly collected information. We have begun with unusually explicit accounts of ideas of modern biology that are fairly well known in order to develop in context our newer assertions. In Fleckian terms, we have written this book in a style that we hope will appeal to those who do not yet belong to an already entrenched, restricting thought community.

In our view biology as a whole, and the origin of meiotic sex in particular, is ready for a paradigm shift—a dramatic change of scientific outlook (Kuhn, 1970). We feel the time has come to question severely, and at least consider replacing, bioeconomic "just-so" stories that rationalize the existence of present-day life forms and their adaptations by thinly veiled comparisons with the profit motive. We feel the "value"—if we may borrow the word—of economic terms such as individual *investment* and *cost* is limited in the new world of flowing genetic information. Nonetheless, we are aware that our view of life is itself dependent on a change in our "directed readiness for perception" (Fleck, 1979). Our descriptions of sex, based on new evidence from cell biology and protistology, are drawn from analyses of microbial interactions including bacterial sex and symbiosis. It is, of course, still possible to view sexuality in more traditional, abstract ways. We welcome readers' responses to our presentation.

Although readers with two or three years of college biology will find this book easier to read than will those without a science background, we hope that everyone interested in evolution, especially the evolution of sexual processes, will stay with the story. We have defined nearly every term that might be a source of potential confusion. Our definitions do not always precisely conform to those in all the subfields of biology. Indeed, basic terms such as *mating types, spores, males, fruiting structures, zygospores, sorocarps, cysts,* and *conjugants* have different meanings in microbiology, protozoology, and botany. At the risk of seeming simple, we have tried to be explicit concerning the use of any such terms. The professional biologist will forgive us while the uninitiated reader will, we hope, be grateful. A glossary is provided at the end of the book for easy reference.

Most of the material in this book derives from the biological literature and

owes an enormous debt to the hundreds of individual scientists who participate in generating the new observations and experiments upon which the
changing, complex thought styles of science are built. Many will find the ideas
of this book very new, since until recently attempts to grapple with the elusive
history of sex and present it in a systematic way have been foiled by a great
dearth of pertinent information. Because we have been liberal in our interpretations here, it is especially important that the reader be responsive.
We welcome any corrections or comments that this material provokes. We
believe that the goal of understanding the evolutionary origins and history of
sexuality of life on Earth is a worthy one and that it is a joint one requiring the
efforts of a large number of serious and curious people.

# 1 • WHAT IS LIFE?

## DNA, *Autopoiesis, and the Reproductive Imperative*

### DEFINING LIFE, REPRODUCTION, AND SEX

To understand sex and its origins we must first know about some fundamental properties of life: autopoiesis (self-maintenance), growth, and reproduction. These three defining properties of all life are visible manifestations of a detailed carbon chemistry—and all three are likely to occur in a total absence of sex. Through an immense stretch of evolutionary time—the first three billion years of life—sex was usually not required for self-maintenance, growth, or reproduction. By *sex* we mean a process characteristic of live organisms only: the complex set of phenomena that produces a genetically new individual, an individual that contains genes (genetic material, DNA) from more than a single source. Whether we are discussing viruses or pandas we will identify the sources of the genetic material for the new individual as *parents*. It is crucial to understand that sex as the production of genetically new beings from different parents has nothing necessarily to do with reproduction, often an entirely different process.

Reproduction is an increase in the number of individuals. Whereas *sex* means the mixing of genetic sources, *reproduction* means copying resulting in the creation of additional live beings. Beings can be both new in the sexual sense and additional in the reproductive sense, in which case they are members of a sexually reproducing species. But this need not be the case. Most organisms in the world in fact reproduce asexually, whether they sexually recombine or not. Individuals, depending on the genetic potential of the species in question, may be formed from only a single parent or from as many parental sources of genetic material as are required. The minimal number of parents varies from species to species. The number of genetic sources re-

9

quired to make an individual may range from one asexual parent (species of the genus *Amoebae*), to two sexual parents (all mammals), to many quasi-sexual parents (members of the genus *Acrasia*), and so on.

In the case of the *Acrasia* cellular slime molds, hundreds of individual cells independently living in the soil come together under the influence of a chemical attractant and fuse. The cells unite to form much larger creatures, "slugs," which are just barely visible with the naked eye. Since there is no net increase in the number of individuals after the cellular union forms—but in fact a decrease—no reproduction may be said to have taken place; the production of a slug must nonetheless be viewed as a sexual event occurring among multiple parents. In some species individuals are even produced in reproductive events involving only part of a single parent. A portion of a parent cell, for example, suffices to form an individual, as when a bacterial spore develops by partitioning off inside a "mother cell." These examples serve to show in brief fashion the broad range of ways in which individuals are produced, asexually and sexually, as well as through the familiar mode of sexual reproduction.

For clarity we distinguish the formation of individuals by reproduction from the process of replication. The term *replication* we limit to the copying process of the genetic material, the nucleic acid molecules (DNA and RNA). Ultimately, all acts of organismal reproduction involve replication, the molecular copying process. The understanding of the chemistry of replication, the production of very long chains of nucleic acid from complementary molecules inside cells, is perhaps the most significant intellectual advance of the twentieth century. Thanks to this new understanding we are for the first time in a position to examine realistically the problem of the evolutionary origins of reproduction and sex.

## AUTOPOIESIS

Two processes distinguish living from nonliving matter: autopoiesis and reproduction. The term *autopoiesis* is derived from the Greek roots *auto* ("self") and *poiesis* ("making"). Entities that metabolize, that is, that chemically maintain and perpetuate their identity despite constant environmental perturbations, are considered autopoietic (Varela and Maturana, 1974). The advances of biochemistry and molecular biology now permit us to identify in detail the basis for the autopoietic phenomenon we call "life." Carbon-compound transformations by enzyme proteins provide the basis for metabolism. These transformations go on at all times in autopoietic beings, always fueled by light or some kind of chemical energy. Thus metabolism is the mechanism of autopoiesis. Although the term autopoiesis is consistent with our intuition about "being alive," it does not simply mean "life." It is more precise. Virus particles, for example, when they are at a distance from their host cells, are not autopoietic entities. They show no sign of self-maintaining metabolic reactions. They do not require a continuous source of energy and carbon compounds in order to persist. In the absence of renewed contact with bacterial, animal, plant, or other autopoietic entities (cells), viruses respond as passively to environmental changes (such as temperature or water) as any chemicals do. Viruses require entry into some autopoietic entity in order for them to reproduce. On the other hand, bacterial spores are autopoietic entities. They exchange gases and other substances with their surroundings as a part of their strategy for self-maintenance.

All autopoietic beings, either single-celled or many-celled, have distinct chemical properties that determine what we perceive as "life." There are no known exceptions. They are all watery. All autopoietic entities are surrounded by membranes that are semipermeable in that, like gate-keepers or customs officials, they permit passage of only certain chemicals. All contain the long "informational" nucleic acid molecule DNA. Within the membrane is the entire complex of nucleic acids (made of discrete units called "nucleotides") and proteins (made of amino acids), along with food molecules, that constitutes the cell. The cell is the minimal unit of both autopoiesis and reproduction. Only one sort of entity besides a cell has ever been known to be autopoietic and reproductive: organisms and their products made of many cells.

The DNA molecule, like the lines of prose in a how-to manual, comprises the information that determines the nature of the organism. The DNA, like prose, is "read" for instructional value, but it is read in manageable lengths. RNA molecules, only slightly different in chemical design from DNA, are involved in the reading process. Because the RNA molecules retain the sequence of information originally in the DNA as the RNA is made, it is said that DNA is "transcribed" into RNA. That is, RNA molecules are made that correspond to one of these manageable lengths of DNA. The sequence of nucleotides in the RNA then directs the linking of amino acids to form enzymes, which have many different functions, including the synthesis of more nucleic acid. Lengths of DNA are transcribed into RNA and then translated into protein. The pieces are just long enough to determine the length of a chain of amino acids sufficient to make a functional enzyme, one type of protein.

These pieces of DNA were known in another context for many years before their chemistry was revealed. They were hypothesized to exist in order to explain the inheritance of certain traits, such as yellowness and greenness in pea seeds. They were called "genes." We now know that genes, originally hypothetical factors determining the presence of discernible traits, can actually be tracked from generation to generation. In short, genes are pieces of DNA that are large enough to have discrete functions. All the DNA in a cell is spoken of as the total genetic material of the cell or as that cell's *genome*.

Apparently the very existence of an autopoietic being depends on the presence of an extremely long and skinny molecule, or several such molecules, of DNA. All cells harbor thousands of manageable lengths of DNA hooked together. DNA determines the sequences of nucleotides in RNA, which in turn determine the fabrication of functioning proteins. In a cyclical way, some of these proteins act specifically to make more DNA and RNA. The proteins made by the reading of DNA by RNA themselves hook together the nucleotides into more DNA and RNA. Certain rules of this process are inviolable: new copies, replicas of a cell's DNA, must be made every time that cell grows to double its size just before dividing to form two cells.

We do not know of anything in this world capable of reproduction that is not autopoietic. Neither do we know of anything autopoietic that is not made of water and a complex variety of carbon compounds, including nucleic acids and proteins. Indeed, nothing less complex than a cell is autopoietic and reproduces.

There is a vast difference between chemical or even biochemical reactions on the one hand and the metabolic sort of biochemical reactions that are the basis of autopoiesis. Even though test-tube biochemistry involves proteins and other macromolecules that are normal components of autopoietic systems, the biochemical reactions by themselves are not autopoietic. Neither are any of the most elaborate and complex chemical systems not based on carbon ever autopoietic. The highest priority for any autopoietic system is to continue to exist. In so doing autopoietic systems have profound effects on their surroundings. Changes in the surroundings then prompt strategic responses by the autopoietic entities. Autopoietic entities have genetic continuity. Chemicals, on the other hand, even macromolecules, are indifferent to existence: chemical systems have no priorities, nor have they genetic continuity.

These differences between autopoietic and other chemical systems are not always obvious, especially when the autopoietic entities involved are microorganisms (microbes). For example, the polishing of silver and the production of plastic are chemical reactions, whereas the treatment of sewage and the production of beer involve autopoiesis. Sewage treatment and beer making require the metabolism and growth of organisms. If sewage is observed micro-scopically, thousands of millions of bacteria of various kinds can be seen. Examination of beer in the making reveals millions of yeast cells per milliliter of fluid. The difference between chemical processes alone and metabolic autopoietic processes can be illustrated by this example. If vats of sewage and beer were rendered sterile by killing all the microorganisms inside them with heat or ultraviolet radiation, some chemical changes might continue for a while, but there would be no beer and the raw sewage would persist unprocessed. What we perceive as the purification of sewage is the autopoietic activities of fermenting, methane-producing (methanogenic) and other bacteria. What we enjoy as the brewing of beer is fundamentally the autopoiesis of brewer's yeast under warm, wet conditions.

Autopoiesis occurs, then, to maintain an organism during its own life, but by itself autopoiesis does not guarantee that an organism will show genetic continuity or that the characteristics of any given organism will persist faithfully through time. The process that ensures genetic continuity is reproduction. But autopoiesis remains the primary process. On the one hand, without it the organism would not survive to reach the stage at which reproduction becomes feasible. On the other hand, autopoiesis does not depend on reproduction, at least within a single generation. An infertile but healthy person with muscle, circulatory, excretory, and other organ systems in excellent running order is autopoietic even though he is unable to reproduce. In an evolutionary sense, however, such an individual has forfeited genetic continuity; he is already dead. At least in today's world, such a person's contact with posterity (barring future arts of cloning) has been irreversibly severed.

The recursive, self-referential nature of autopoiesis was probably already installed in the first entities on Earth capable of reproduction—the first cell-like beings. If it was not, then the original environment of Earth was spectacularly supportive of naked genes, providing them with at least twenty amino acids, five nucleotides, and just the proper concentrations of salts and other molecules they needed to reproduce. Reproduction depends on autopoiesis, but autopoiesis in turn cannot take place without the eventual replication of DNA. An inheritance of biotechnology from the parent is required if an organism is to acquire or produce food molecules, nucleic-acid-fabricating enzymes, and so forth. Within a cell, DNA in the presence of small molecules (familiar to us as "food" or "nutrients") and specific enzymes called "polymerases" forms a second, identical DNA molecule. This property of DNA copying, what we here call "replication," is the basis of all autopoietic cell reproduction. The familiar process of growth in size of any organism depends intimately on the DNA copying process. Inside the cells of the growing organism DNA replicates, RNA containing the information stored in that

DNA is produced, and from that RNA new proteins are formed. The cell takes in food and processes it according to the instructions in its DNA and RNA, it doubles in volume and finally divides. The single DNA molecule has become two by replication; the single cell has doubled via reproduction.

## AUTOPOIESIS AND REPRODUCTION AS IMPERATIVES

Autopoiesis is not an option for a living being—it is an imperative. It is absolutely required at all times for any member of the biota. The biota has been defined as the sum of all life on Earth. Once autopoiesis appeared in the events that marked the origin of life on Earth, it was never lost. If the self-maintaining system of a cell fails, the cell dies. If the cell is a component of a larger organism capable of replacing a dead cell or group of cells, the organism in question persists as an autopoietic entity. If too many cells or groups of cells lose their capacity for autopoiesis, death of the organism ensues. Even though we do not necessarily detect directly such chemical transformations as nutrient uptake; DNA, RNA, and protein fabrication; and energy conversions, they occur rapidly and continuously in every autopoietic entity. Autopoiesis is also an absolute prerequisite to reproduction.

Perhaps it is because autopoietic systems are always replicating their DNA and making more of the other large molecules that at some point they tend to reproduce. Both bacteria, after reaching a critical volume, and cells with nuclei, after reaching a certain value of the ratio of the volume of the nucleus to that of the cytoplasm, tend to divide. Growth is accompanied, eventually, by reproduction. Tiny bacterial cells may reproduce in fewer than fifteen minutes. Elephants and whales may require twenty years. Whatever the rate, the mechanisms of reproduction always involve DNA replication. In addition they may involve cell growth, cell division, embryological development, and other processes. In principle it is not clear why autopoietic entities are divided into recognizable individuals that reproduce; this simply seems to be the case. Reproduction of cells seems to have evolved when bacterial cells themselves evolved; it seems to be a biological imperative.

Sex, on the other hand, is optional. It is a whole series of processes that vary enormously in detail. Sex, the formation of new individuals having more than one parent, is not required at all for either autopoiesis or reproduction. The confusion results from the fact that sex has become as natural a part of certain lineages as are clothes to modern humans. We take both for granted, but both can be shed. It is important to remember the true state of cellular affairs regarding sex: autopoiesis, when stripped to its bare essentials, reveals that sex is superfluous, an almost awkward adjunct to reproduction. It is only the peculiar history of animal evolution that so far has prevented human beings and many other animals from "budding"—reproducing asexually after the fashion of tree cuttings.

We shall see how, in some organisms belonging to only a very few phyletic lineages (family trees), a certain kind of two-parent sex became intimately connected to reproduction because of events occurring in ancestral phyla. In some of these groups of organisms sex came to be required for continuing autopoiesis and reproduction. The fact that we are descendants of that minority of organisms that is sexually reproducing has led us to a very unbalanced and unrealistic view of biological sexuality. Instead, we should begin our narrative with the very first sign of sexuality in the history of life and trace the story of the development of the different forms of sexuality in autopoietic beings. Following this tale should help us restore our balance and perspective and provide for us a biological context for our own sexuality.

# 2 · WHAT IS EVOLUTION?

## Cell DNA Continuity and the Necessity of Error

### HOW REPLICATION WORKS

This book attempts to tell the story of the evolution of sex. But before we can discuss the relationship between sexuality and evolution we must know, in biochemical terms, what evolution really is. To know that, we must first know what reproduction is, because it is only through *differential reproduction*—the reproduction of some organisms more than others—that evolution, or changes in populations over time, can occur. Ultimately, all acts of reproduction, from bacterial parasites bursting from their host cells to the birth of a human infant, are based on the replicative ability of DNA.

Understanding, in chemical detail, the act of DNA replication is the real triumph of modern biology. In nature the smallest unit capable of replication is a nucleic acid sequence called a "small replicon." Examples of small replicons include viruses, plasmids, episomes, and other entities we will discuss.

To grasp the simplicity, elegance, and clarity of the replication process, it is necessary to know only the rule of the complementary base pairs that make up DNA. DNA, the long, doubly coiled molecule that contains the genetic information both to construct intricate living organisms and to make more of itself, is composed of four kinds of flat bases arranged like steps on a ladder (fig. 1). These flat bases are carbon molecules with some hydrogen, oxygen, and nitrogen atoms in them. The order of the four bases—molecular letters in a nimble four-letter alphabet—is crucial, as the sequence of the bases determines the uniqueness of each individual. The total number of bases making up the DNA double helices is equivalent to the total quantity of genetic material comprising an individual's genome.

Replication, the basis for that strange biological equation $1 = 2$, can be

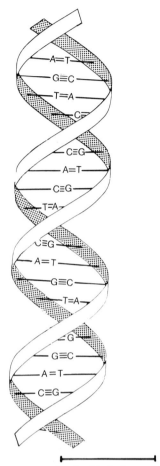

**Fig. 1.** DNA structure. The double helix is a long mole-cule composed of four bases (smaller molecules) arranged like steps on a ladder (bar = 10 Ångstroms). (Drawing by Steven Alexander.)

performed only by nucleic acids, that is, either DNA or RNA. Once the rule of complementary base pairs is understood, it will become clear why replication is limited to the nucleic acids. Discovering the meaning of nucleic acid base pairing not only won the Nobel Prize for its discoverers, James Watson and Francis Crick, but it was fundamental to the entire "molecular biology revolu-tion." Here, in brief, are the crucial pairing rules, the basis of the replication of molecules and therefore of the reproductive ability of all life.

Below are the four types of DNA bases and their respective abbreviations:

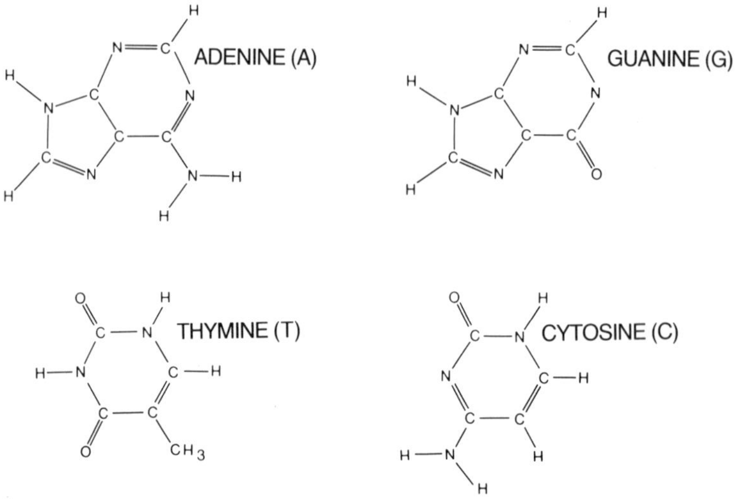

ADENINE (A)

GUANINE (G)

THYMINE (T)

CYTOSINE (C)

Adenine and guanine, collectively known as purines, are larger molecules than thymine and cytosine, collectively known as pyrimidines. Each of the four bases differs only slightly in chemical structure from any other. These structural differences carry biological information, just as the small structural difference between the letters c and o carries cultural information. In order to fit as steps across the double helix "ladder," adenine pairs with thymine (A–T) and guanine pairs with cytosine (G–C) (see figure below). (The dotted lines denote hydrogen bonds, two in A–T and three in G–C.) Adenine-thymine (A–T), then, forms one base pair and guanine-cytosine (G–C) another. Enzyme proteins break the rungs that join the two strands of the DNA. A new strand is formed as a complement to each of the originals. That is, opposite each thymine a new adenine forms, and opposite each cytosine a new guanine forms. Thus, because A's couple only with T's and G's only with C's, new A–T and C–G pairs are joined by newly formed rungs. What before replication was a single double helix of DNA a millimeter long is now, after replication, two double helices, each one a millimeter long. The schematic doubling of a portion of the DNA of any organism is shown in figure 2. We can, of course, show only a snippet; even a virus would have far more DNA than can be displayed here. Cells have at least 5,000 genes coding for 5,000 proteins, and about 200,000 nucleotide pairs are required to make each protein. Thus even a small cell has about a billion nucleotide pairs.

ADENINE  THYMINE

GUANINE  CYTOSINE

The Watson-Crick discovery is that the structure of DNA itself immediately suggests how information is stored in autopoietic beings and how replication works. The principle by which DNA doubles governs the pattern of RNA synthesis as well. The sequence of nucleotides in a strand of DNA dictates a complementary sequence in RNA. Nucleotides contain—in addition to the bases A, T, G, and C—sugars and linked phosphates, which form the structural "backbone" of nucleic acid. RNA directs the fabrication of proteins. Some of these proteins accelerate the tendency of DNA to double again, others act as catalysts or structural components, agents of the cell's metabolism ensuring its autopoietic behavior in the face of continual external threats to the cell's existence. These are universal truths for all known organisms.

## MUTATIONS

If, from the beginning, DNA had behaved as a perfect self-copying machine and all base pair-halves found their complements with 100 percent accuracy, the world would soon have been populated by the reproductive products of a single entity. All living beings would be structurally identical. Life on Earth would consist of a myriad of creatures all exactly like the original

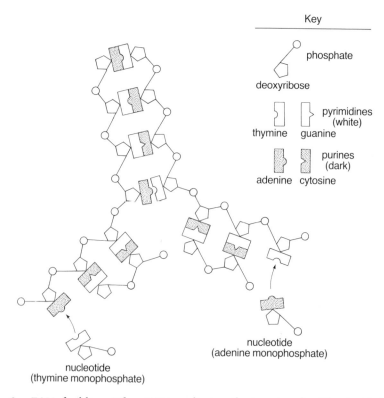

Key

phosphate

deoxyribose

pyrimidines
(white)

thymine    guanine

purines
(dark)

adenine   cytosine

nucleotide
(adenine monophosphate)

nucleotide
(thymine monophosphate)

**Fig. 2.**   DNA doubling. When DNA replicates, the two strands of the double helix separate and nucleotides attach to their complements on the single strands. Because of their sizes and chemical properties, adenine always pairs with thymine and guanine always pairs with cytosine. In this way two new double helices are built up from the two strands of the original double helix. (Drawing by Laszlo Meszoly.)

parent. Such genetically identical products of reproduction, members of a *clone*, would constitute what, in Darwin's terms, might be called "descent without modification." But descent without modification rarely occurs. It does take place in the laboratory, when a single bacterium reproduces on the surface of a small petri plate and grows, after a few days, into a visible clone. But this continues for only a brief spell. Then mutations, errors in the DNA copying process, mar the idealism of a perfect clonal lineage even in the asexual bacteria. If there were no mutations at all, evolution would not have happened and would not be happening now. Even one mistake in every hundred million copies of a functional piece of DNA is enough to fuel evolution. Organisms that are changed through mistakes in DNA copying naturally pass these changes on. Sometimes the changed offspring reproduce at a greater rate than their parents. This is what is meant by *differential reproduction.*

The genetic mutations occurring in parental generations provide the raw material out of which each succeeding generation develops. The stochastic opportunism with which life evolves depends on the tendency of its copying systems to generate "errors." Organisms that bear mutations are mutants. Evolution requires the continual emergence of mutants, which differ from their parent or parents. Sex, because it resembles the process of mutation in its ability to produce discrepancies, has been long assumed to be a major factor in the generation of evolutionary novelty. The noteworthy biologist Gavin de Beer (1972) even went so far as to write that "If it had not been for exchange of genetic material (i.e., sexual reproduction), there would have been no adaptation or evolution at all." This statement has not been proved to our satisfaction, nor, despite its constant reiteration in the biological literature (Ghiselin, 1974; Bell, 1982), is there any solid factual evidence to back it up. In fact animal and plant sex, although it clearly produces novel organisms in each generation, probably has very little to do with the direct generation of lasting *evolutionary* novelty. This is because the novel types that are created by the conjunction of disparate types of the same species may lose the very novelty they have gained when they reproduce and thus remix their genes. Sex is as much a sink of variety as it is a source and thus effectively cancels itself out in the evolutionary long run. Parthenogenetic populations of lizards and rotifers, for example, show as much variation as comparable populations of sexual lizards and rotifers (Buss, 1983a; Bell, 1982).

Evolution, the changes of organisms through time, is a process with at least three components: reproduction, mutation, and differential survival (natural selection). Faithful replication of DNA, accompanied by its entourage of enzymes, forms the basis of reproduction that produces organisms like their parents. All populations of organisms have a tendency to increase in number exponentially. This tendency, measured in units of organisms produced per unit of time, is called the *biotic potential.* For example, the biotic potential of people is about ten children per generation of twenty-five years, that of dogs about a hundred puppies per year, and that of the colon bacterium about a billion per day. Certainly the fact that a species has a high biotic potential does not mean that it has a high standing population. Passenger pigeons have a higher biotic potential than that of humans, for example, but their numbers are far fewer.

Some populations show lower rates of mutation than others. Mutator genes have even been found that influence the rate of mutation (Suzuki, Griffiths, and Lewontin, 1981). Mutants, by definition, pass on their changed genes to their offspring. No matter what the rate of production of mutations in any population, *some* organisms are mutant and thus differ from their fellows. This means that at any time, in any place, different organisms exist to vie for the limited resources of the same planet, or forest, or droplet of nutrient-rich water. Since mutant organisms are different from their forerunners, they tend to enjoy advantages or suffer from disadvantages relative to their predecessors and neighbors. Their differences are reflected in differential rates of re-production. Eventually certain lineages of organisms overwhelm others: house mice have overwhelmed island voles on Manhattan Island. First local populations, then entire species become extinct under the duress of competi-tion, inadequate resources, and other factors that absolutely limit biotic potential.

Spontaneous deletions, additions, and alterations of DNA base pairs are the kinds of mutations primarily responsible for the generation of new variety. Mutations in laboratory rats, for example, do not simply result from exposure of the rodents to chemicals, radiation, or heat. Laboratory rats develop muta-tions whether or not they are given treatments known to induce mutations. Indeed, mutations are spontaneous and natural in all members of all popula-tions. In people, heritable changes—the appearance of new mutants—are difficult to observe. One needs information ruling out the possibility that the change is environmentally induced, that it is not just a physiological change that will never outlive its generation. In order to qualify as mutations, that is, as true errors in genetic fidelity, changes must be passed on to offspring.

To verify mutations is comparatively easy in bacteria, some of which reproduce three times an hour. It is easiest to identify a mutation in humans when it is an obvious embryological one, such as the presence of additional fingers (polydactyly). This sort of trait can be traced as it is passed on from parents to children. Also, because mutations in genetic material cause corre-lated changes in proteins, the study of such protein changes can guide the researcher toward detecting underlying alterations of nucleic acid base pairs. Currently a very precise method of detecting mutations is by chemical study of the amino acid sequence in purified, well-known proteins. A direct method of study of complete protein sequence and structure, the most precise known, uses an X-ray diffractometer. The process is laborious, but by studying the deflection of X rays by crystalline proteins onto photographic film (or light-sensitive screens) one can map the exact chemical makeup of a given protein's amino acid sequence and its folding into a three-dimensional structure.

By identifying the complete amino acid sequence in a protein it is possible to deduce, from the genetic code, the most probable DNA nucleotide base pair sequence to have determined the protein. Likewise, if one experimen-tally isolates and studies the complete nucleotide sequence of a stretch of

DNA, one can specify, from the rules of the genetic code, the protein structure that the piece of DNA determines. A surprising finding of the new molecular biology is that there does not exist a one-to-one correspondence between the DNA nucleotide sequence in a cell's genome and the sequence of amino acids in proteins. In general, it takes three nucleotide bases, called a "triplet," to specify a single amino acid. This run of three base pairs is called a "codon." Furthermore, several different groups of nucleotides can specify the same amino acid. (For example, the amino acid phenylalanine can be specified by the DNA nucleotide sequence TTT or TTA.) There are plenty of triplets of bases to code for the needed sequences, since there are only twenty different amino acids making up all proteins, and the number of different three-letter nucleotide sequences is four (the number of different bases) raised to the third power (the number of bases determining an amino acid), or sixty-four.

More spectacular still is the observation that, at least in the case of animal and plant cells, hundreds of thousands of DNA base pairs do not specify any proteins at all. There is no accepted explanation for this observation. Indeed some, such as Doolittle and Sapienza (1980), have argued that this "extra DNA" is nonfunctional, that it is "selfish DNA" (Dawkins, 1976) that is in the cell simply because it is intrinsically capable of replication and therefore can stick around. We feel that it is far more likely that the large quantities of "extra" DNA (nonprotein-coding DNA) have an evolutionary explanation. We shall discuss the possibility that at least some of the noncoding DNA in plant and animal cells is a consequence of the history of the sexual systems in protists, the ancestors of animal and plant cells.

Mutations, defined broadly as heritable changes, undoubtedly are the means of generating the novelty upon which evolution depends, but mutations occur at different levels. Not all mutations are single base-pair changes in DNA; some, for example, can appear as changes in entire genes. Genes, as units of function, can be duplicated—all of the 200,000 or so nucleotide base pairs making up a gene can be repeated. Genes make products of importance to the cells they help comprise. Many genes, acting together, determine observable traits. A trait or inherited characteristic of clear importance to the cell or organism that bears it is a *seme*. A seme is always determined by more than a single gene and its single protein product. The appearance and change of semes is of crucial importance to the evolutionist. Fundamental semes include metabolic pathways, such as Embden-Meyerhof glucose breakdown in cells; flowers, cones, and leaves in plants; and gill arches, swim bladders, lungs, teeth, uteruses, and hearts in vertebrates. A single mutation cannot produce a seme, although it can change or destroy one.

Sexuality cannot produce a seme either, although it can destroy or change one by rearranging the genes that determine it. The changes produced by sexuality are in a completely different category from mutations. They are really temporary rearrangements of genes. Because variety produced by sex is often nullified by further sex, such rearrangements are not permanent. Genes produce variation by many different means. Organisms always seem to have a basic reservoir of standing variation. Strong and directed selection pressures, not increases in amounts of variation, lead to evolutionary changes. Both permanent mutations and temporary rearrangements are the bases for the variety of living forms we see around us. But beneath the diversity of form and at the basis of the inherited changes lies an extraordinary unity of macromolecular chemistry. The diversity ensures that some organisms will differentially leave more offspring; natural selection therefore will lead to evolutionary change. The chemical unity of DNA replication and protein fabrication ensures high heritability of traits and the continuity through time of ancestors and descendants. It is just this genetic continuity through time, as Darwin so perceptively recorded, that indicates to us that all organisms on Earth are ultimately descended from a common stock (Darwin, 1859). We now know that the ultimate common ancestor was a DNA-containing microbial cell. It is this incredible chemical and cellular conservatism that permits us to study live organisms and from them deduce the parts of the puzzle, putting together plausible answers to the riddle of their sexual origins.

# 3 · WHAT IS SEX?

## Molecular, Cellular, and Organismal Recombinations and Fusions

COMING TOGETHER

Whereas sexual processes must involve the production of genetically new individuals, they do not require an increase in the total number of individuals. Reproduction, on the other hand, invariably results in an increase in the number of live beings. The minimal number of individuals before an act of reproduction is one; the minimal number afterward is two. One individual may, in a single act of reproduction, form only one more individual. This is the usual case when a bacterium or an amoeba divides. A single individual, in other instances, may form up to several hundred offspring in one reproductive event—for example, when a *Bdellovibrio* bacterium enters and reproduces inside a host cell; after several hours, twenty or thirty newly formed *Bdellovibrio* bacteria burst out.

Autopoiesis is an absolute prerequisite for reproduction, but not for replication. No small replicons such as DNA or RNA viruses, as we have noted, are autopoietic entities. Neither are isolated DNA or RNA molecules in a test tube autopoietic. Yet all of these are capable, given adequate chemical conditions, of replication: the production of complementary copies requires only the chemical precursors (nucleotides), the correct polymerase enzymes, and permissible temperatures, salt concentrations, and other appropriate ambient conditions. However, although certain nonautopoietic entities can replicate, only autopoietic entities can reproduce. Even though a single viral reproductive act may produce two hundred viruses from one parental virus particle, viruses are totally incapable of reproduction in the absence of an autopoietic host (namely, an appropriate kind of bacterial cell for a bacterial virus or a plant or animal cell for a plant or animal virus).

A sexual event, whether in viruses, bacteria, protists, fungi, animals, or plants, absolutely requires that at least one of the parents be an autopoietic entity. This generalization correctly implies that sexuality involving viruses in the absence of host cells is not possible; this has, of course, never been observed.

Fungi probably hold the record for reproductive capacity, for a high and sustained biotic potential. Thousands of spores (conidia) can form and be released in a few hours from the "heads" (conidiophores) of many fungi, for example, from the head of the common woodland fungus *Alternaria*, shown in figure 3. A single amoeba may partition itself into several smaller amoebae at one time. This sort of multiple reproduction, often labeled "multiple fission," occurs, for example, inside the cysts, or hard-walled protective structures of *Pneumocystis carinni*. *Pneumocystis* is an enigmatic microorganism often found in lung tissue of people with poor general health or reduced resistance to bacterial infection. It is nearly invariably associated with patients who have autoimmune deficiency syndrome (AIDS).

A very dramatic and well-known example of multiple reproduction occurs when malarial parasites, organisms belonging to the genus *Plasmodium*, multiply inside the red blood cells of afflicted persons. A single microbe, called a "trophozoite," attaches to a red blood cell and forcibly enters it through a structure of its own making. Once inside, the *Plasmodium* rapidly makes many copies of its own DNA inside its nuclei, using human hemoglobin as its major food. The original nucleus divides several times, and the

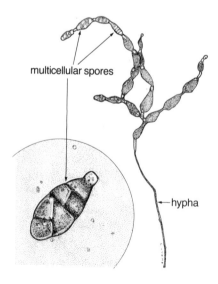

multicellular spores

hypha

Fig. 3. *Alternaria* spores. The tips of *Alternaria tenuis* stalks (top) in a few hours can produce thousands of spores (bottom), which upon release ensure the high biotic potential of this woodland fungus. (Drawing by Christie Lyons.)

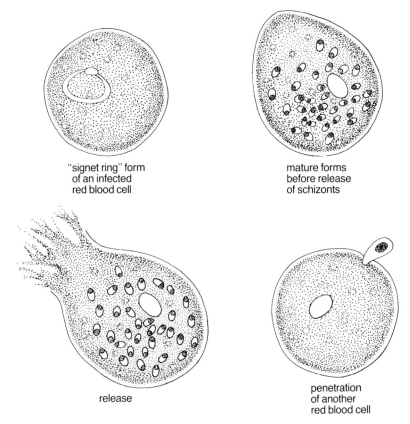

"signet ring" form
of an infected
red blood cell

mature forms
before release
of schizonts

release

penetration
of another
red blood cell

**Fig. 4.** Multiple reproduction and release of the malarial parasite *Plasmodium*. The parasite feeds on the hemoglobin in the red blood cells of infected persons as DNA is replicated in its nucleus. Release of the schizonts apparently happens quickly and is very difficult to observe directly. (Drawing by Christie Lyons.)

cytoplasm cleaves correspondingly to form secondary trophozoites. After a few hours up to hundreds of secondary trophozoites burst out and, in doing so, break open the blood cell. They are thus abruptly released into the blood fluid. They float until they reach and attach to new red blood cells, where the process is repeated. Malarial victims often have periodic fevers, which are directly caused by the release of these trophozoite offspring, produced by multiple cell divisions, into the blood (fig. 4).

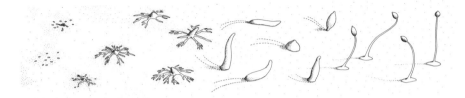

**Fig. 5.** Fusion of *Acrasia* cells. It takes hundreds of *Acrasia* slime mold cells to form one slug. The fusion of these cells is an instance of a sexual process resulting in a reduction of the numbers of individuals. The figures at the far left did not fuse; in this case the individual amoebae were repelled from each other. (Drawing by Christie Lyons.)

Since all these examples involve an increase in the number of individuals, they are all descriptions of reproduction. In none of the examples (bacterial and amoebic division, conidia formation in fungi, *Pneumocystis* and *Plasmodium* growth) are genetically new, that is, recombinant, offspring produced. None of these organisms has more than a single parent. Thus all, in spite of their profound differences, are products of *asexual reproduction*. The number of organisms has increased in every case.

In a sexual process the number of individuals does not necessarily increase. Indeed it may decrease. This is the case, for example, when several hundred *Acrasia* cells fuse to form one slug (fig. 5). We also know that the number of individuals may simply remain the same during or just after a sexual event, as, for example, when two paramecia conjugate, exchanging nuclei (fig. 6). Each member of the conjugating pair passes and receives new DNA inside the nuclei, but neither produces additional offspring in the conjugation process. For any process to be considered sexual there must be a coming together and joining of genetic material from two sources in a single individual. It is this union to which we refer, throughout this book, as the sexual process or sexual act.

The joining in sexuality may occur at any of several levels, which may be present either singly or together. We identify these levels, roughly in increasing order of size and complexity, in table 1.

We shall deal with examples of each of these cases in turn, in roughly the order in which they evolved. In the simplest kinds of sexuality, which first occurred among viruses and bacteria, DNA molecules from different sources

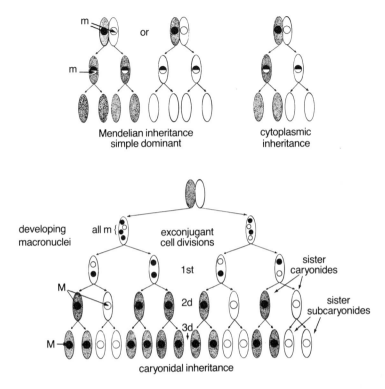

**Fig. 6.** Inheritance patterns in *Paramecium*. M = macronucleus; m = micronucleus. (Drawing by Christie Lyons.)

(parental DNA) separate and reunite in a different combination, as when a virus enters a cell. This molecular transfer from the autopoietic entity from which the DNA was formed to a second autopoietic entity is a kind of recombination.

The uniting of strands of DNA double helices to form a genetically different molecule, part of which is derived from one parent and part from another, is an instance of *genetic recombination*. The product of such a molecular-level union, a DNA molecule with an altered nucleotide base pair sequence, is called "recombinant DNA." Since 1975, when the laboratory techniques were first developed, however, most biologists and biochemists reserve the

**Table 1.** Levels of Sexual Union

| LEVELS OF UNION | EXAMPLES |
|---|---|
| Genophoric or chromonemal | DNA recombination, small replicon entry into host cells, integration into host DNA |
| Chromosomal | Chromatidal: crossing over |
| Nuclear | Fusion of nuclei: karyogamy |
| Cellular (conjugation) | Fusion of cytoplasm: syngamy, fertilization |
| Histological | Transplantation of isolated tissues or organs, pollination, egg or sperm transfer |
| Organismal | Fusion of entire organisms or their parts: mating or conjugation, gamontogamy |

term *recombinant DNA* for the artificial union of DNA, that is, the fusion of DNAs from different origins achieved by human intervention. Whether natural or artificial, the union of such DNA uses specific DNA-splicing and patching enzymes (nucleases, ligases). Consistent with this usage, we refer here to natural genetic recombination simply as recombination, or sex. Artificial or man-made DNA sequences we distinguish as such.

But not all sexual events necessarily involve a molecular fusion, that is, the formation of a new DNA molecule. Recombination in the sense of rearrangement can occur in a sexual process without actual fusion. What else, then, may come together? Nuclei with chromosomes, cytoplasm, entire cells, parts of organisms, and entire organisms; recombination can occur or be halted at all these levels. One aspect of the confusion surrounding the problem of the origin of sex stems from a failure to recognize the multiplicity of sexual phenomena and the diversity of sexual processes. Although in all cases of sexuality a new genetic combination forms that differs from its parents, the nature of that new form and the level of newness depend on many factors. The major factor is, of course, the species or kind of organism and its biological potential.

## DISTINCTIONS AMONG BACTERIAL SEX, MEIOTIC SEX, AND PARASEXUALITY

Before presenting examples of coming together at various levels, we should clarify the major kinds of sexuality. There are only two: prokaryotic recombination and eukaryotic meiotic sexuality. Prokaryotic sex generally involves a fusion on the molecular level, that is, DNA recombination. Meiotic sex involves reduction of chromosome numbers inside membrane-bounded nuclei and subsequent fertilization, which by fusion of nuclei reestablishes these numbers.

All life can be divided into two major styles of cellular organization: the prokaryotic, including all bacteria, which lack membrane-bounded nuclei; and the eukaryotic, which includes all organisms with nuclei. Some groups of organisms of formerly uncertain status, such as the "actinomycetes" (now called the "actinobacteria") and the "blue-green algae" (now "cyanobacteria"), are clearly prokaryotic and are therefore placed in the kingdom Monera. Monerans, by definition, lack membrane-bounded nuclei and therefore include all prokaryotic cells and organisms composed of prokaryotic cells. The kingdom Monera contains at least 10,000 kinds of organisms, all of them microbial.

Although viruses are not autopoietic entities and therefore are not members of the kingdom Monera, they do engage in DNA recombination, the prokaryotic style of sex. At present this kind of sexuality is well known for many viruses, but it always occurs inside the cells of their hosts. No matter what type of sex the host undergoes, viruses retain the prokaryotic type of molecular-level DNA recombination. Even though many viruses and bacteria are known to be capable of genetic recombination, sexuality in prokaryotes was discovered only in 1946. When he was only nineteen years old Joshua Lederberg, now the president of Rockefeller University, followed up earlier experiments in which a harmless variety of pneumococcus bacteria was transformed into a disease-causing agent through contact with heat-killed virulent bacteria or extracts from the heat-killed bacteria (Avery, Macleod, and McCarty, 1944; see also McCarty, 1985). Lederberg was able to prove definitively that bacteria have a sex life, that genetically determined traits, such as antibiotic resistance or ability to use lactose as a source of food, are passed from donor to recipient bacteria (Lederberg, 1955).

*Meiotic sexuality*, on the other hand, had been discovered by the end of the nineteenth century, when pollen-tube germination and growth toward the egg in the ovary of flowering plants was seen by botanists and fertilization of eggs by sperm was observed by zoologists in many species of animals.

Because most discoveries of sexuality in nature can be explained on the principles of either prokaryotic DNA recombinations or eukaryotic variations on the meiotic theme, "comings together" that are neither of these two things are usually refered to as *parasexual*, a term that we will retain. Parasexuality will be defined as any process, exclusive of both "standard" prokaryotic recombination and meiotic sex, that brings together separate genomes in a single individual. The term was first applied to fusion of fungal strands (hyphae) to make a new recombinant organism with two kinds of nuclei (a heterokaryon) followed by a later reduction to only one kind of new nucleus in a process that did not involve standard fertilization and meiosis (Pontecorvo, 1958). Given our broad perspective, however, the process of fusion of different and once separate amoeba cells in the formation of the *Acrasia* slug is also an example of parasexuality. Parasexuality, more than standard sexuality, tends to be an occasional and irregular process, differing widely in the various species that display the phenomenon.

## LEVELS OF FUSION

Bacteriologists and molecular biologists often loosely speak of the bacterial chromosome. Bacterial DNA is, however, so different in structure from the true chromosomes of eukaryotes that failure to recognize this difference has precluded communication between scientists and their students and has led to great confusion. The prokaryotic and viral genophore (gene-bearing structure) must be clearly distinguished from the eukaryotic chromosome (Ris, 1961) in order for the story of the evolution of these structures to be told with any authenticity.

Chromosomes were first recognized under microscopes by cytologists. After a treatment with strong acid called "acid hydrolysis," the deoxyribose sugar of the DNA can be prepared to stain bright red with the Feulgen reagent, named for the German cytologist Robert Feulgen, who was active in the 1920s. A carbon-nitrogen double bond is formed between the stain molecule and the stacked sugar residues on the DNA. The bright red chromosomes can be seen, and in favorable cases even their numbers can be counted. The Feulgen reaction made possible hundreds of studies of chromosome behavior in eukaryotes. We now know that chromosomes are composed of DNA having several different kinds of proteins tightly wound about it. By weight chromosomes are about 60 percent protein, 40 percent DNA.

The prokaryotic genophore, unlike the chromosome, consists mainly of linearly ordered, membrane-attached, naked DNA. It is far less complex and far narrower in function than the chromosome. Because of its different structural organization, it stains less well with the Feulgen reaction. The fundamental strand of the DNA of the prokaryotic genophore is only 25 angstroms in width. (An angstrom is a convenient measure of length for chemical bonds in atoms; see table 2.) The bacterial DNA has been called the *chromonema*, or "colored thread," by analogy with the chromosome ("colored body"), even though it often does not take on Feulgen's colored stain. The fundamental DNA-protein strand of the eukaryotic chromosome is far wider, some 100 angstroms when prepared under the same conditions for visualization as the chromoneme. (DNA structures are very sensitive to salt concentration, temperature, and other conditions of preparation; therefore these must be specified when appearance is mentioned.)

At least during some part of the life cycle of eukaryotic cells, they invariably have a number of chromosomes, each composed of tightly coiled 100-angstrom-long protein-coated DNA strands. The continuous threadlike material is called "chromatin." The tight coils of the protein around the DNA produce a structure of chromatin like that of tiny beads (nucleosomes) on a string. These beads have an octagonal symmetry and can be visualized with very high powered electron microscopes. Nucleosomes are unknown in prokaryotic genophores, another reason that the word *chromosome* ought never to be used for bacterial structures.

## CHROMONEMAL SEXUALITY

The distinctions between genophores of bacteria and chromosomes of eukaryotes are important in discussions of sexual fusion because these structures are involved in the recombination events. Bacterial and viral recombination, regardless of the organismal fusing events that precede it, always involve chromonemal (and never chromosomal) processes. Such prokaryotic

**Table 2.** Units of Measurement

| Unit | Meter | Useful for Measuring |
|---|---|---|
| Angstrom | $10^{-10}$ | Macromolecules; DNA widths |
| Nanometer (≡millimicron) | $10^{-9}$ | Cell ultrastructures: flagella, microtubules, chromatin, kinetochores, centrioles, width of nuclear membrane |
| Micrometer (≡micron) | $10^{-6}$ | Bacterial cells; chromosomes; undulipodia; protists |
| Millimeter | $10^{-3}$ | Eggs; egg sacs; protoctists; bacterial colonies; sperm sacs |
| Centimeter | $10^{-2}$ | Mammalian sex organs; fungal sexual structures (mushrooms) |
| Meter | 1 | Animals; plants |

recombinations are of two kinds. In one process a genome or a part of a genome is acquired by a prokaryotic cell but not integrated into the linear order of the cell's DNA. In the other the acquired genome or partial genome is integrated into that order.

In the first case the extra piece of DNA is replicated and carried along independently of the rest of the genophore. The extra DNA, which is smaller than the genophore, may replicate faster, with the result that cells having supernumerary DNA fragments are formed. It may replicate more slowly than the genophore, in which case cells lacking the piece tend to be produced. Because these unintegrated genetic elements, small pieces of DNA, were detected with different techniques and have different origins, they have many names, including small replicons, plasmids, F-factors (fertility factors), viral DNA, and episomes (Sonea and Panisset, 1983).

A resulting bacterial cell that contains its own original genophore and an additional piece of unintegrated bacterial DNA has been called, by analogy with the act of fertilization, a "merozygote" (partial zygote, the cell formed by the junction of two gametes, or sexual cells). All these examples of small genomes exchanged by bacteria can be thought of as stages in the development of viral-mediated bacterial sex, as indicated in figure 7 (De Long, 1983). The mature virus particle, an example of which is shown in figure 8, does not have to be formed in order for recombination to occur. It is simply far more resistant to environmental degradation than DNA in solution.

In the event of chromonemal sex, the DNA of the second parent may be integrated into the linear order of the genophore. This is the usual form of bacterial conjugation. The mechanism by which the new DNA forms is the process of DNA recombination, which always involves a complex interaction of different enzymes with the parental nucleic acids. For the resulting cell to be viable it must have a complete set of genomal DNA, that is, at least one copy of each essential gene. Generally, since the second parent will bring in alternative genes to those the first parent already has, the recombinant cell must rid itself of excess gene copies.

Biochemically, then, bacterial conjugation is characterized by a series of events. First, two potential parents must make contact. Second, the parents must remain attached to each other long enough for the donor cell to pass its DNA through its plasma membrane and cell wall and into the recipient cell. This DNA must align with the recipient DNA accurately enough to assure that a recombination event does not duplicate or eliminate a base pair. This requires exquisite recognition of base-pair sequences by enzymes that bind to specific sequences. The recipient DNA must be spliced and excised, the donor DNA must be inserted, and the recombinant DNA must be patched, or

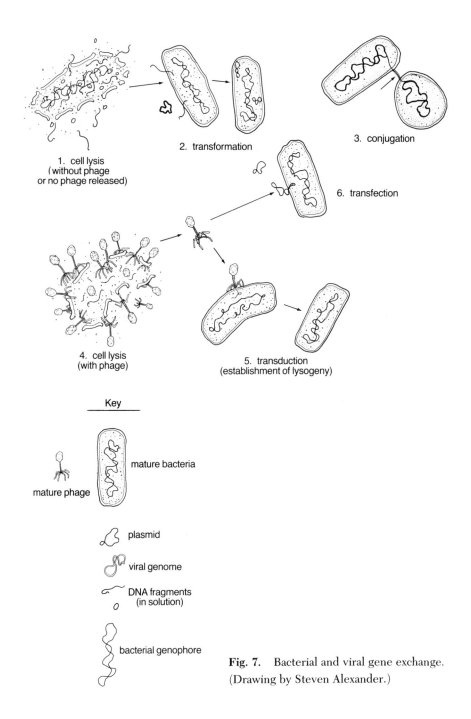

2. transformation

3. conjugation

1. cell lysis
(without phage
or no phage released)

6. transfection

4. cell lysis
(with phage)

5. transduction
(establishment of lysogeny)

Key

mature bacteria

mature phage

plasmid

viral genome

DNA fragments
(in solution)

bacterial genophore

**Fig. 7.** Bacterial and viral gene exchange.
(Drawing by Steven Alexander.)

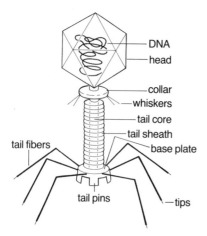

DNA
head
collar
whiskers
tail core
tail sheath
base plate
tail fibers
tail pins
tips

**Fig. 8.** Mature virus particle, a bacteriophage. (Drawing by Laszlo Meszoly.)

"ligated," all by enzymes. Finally, the excess unused DNA of the donor must
be destroyed by digestion with nuclease enzymes. From many genetic experiments it is well known that this process often occurs with no detectable errors.
Regardless of the precise details of the enzymatic processes performing this
genetic surgery, all these elements of the process must occur in each sexual
event to ensure the production of a healthy recombinant. For details of the
recombination process see Radding (1978).

There is no reliable estimate of precisely how many different enzymes are
involved in the bacterial sexual process, but one suspects several dozen at
least. Some have been identified by observing mutant organisms deficient in
these enzymes that cannot conjugate. At any rate, even the simplest bacterial
chromonemal sexual process is, from a biochemical point of view, not at all
simple. This complex sequence of events, all needed to ensure the success of
the final recombinant product, could not have evolved full-blown without
predecessors. Certainly the enzymatic processes listed here did not develop
in order to permit bacterial conjugation to evolve: evolution is never fore-
sighted. In the next chapter we attempt to outline the possible evolution of
the chromonemal sexual processes based on both the known history of early
life and the biochemistry of these processes.

## CHROMOSOMAL AND MORE COMPLEX LEVELS OF FUSION

Other sexual processes involve even more complex structures and bio-
chemistries than do those of prokaryotes. Eukaryotic sexuality always and

intrinsically involves nuclei and chromosomes. In the standard cases the cycle of meiosis and fertilization occurs. The number of chromosomes per nucleus of each eukaryotic cell remains constant and characteristic for the species. This number, sometime during the life cycle of the organism, must be doubled (made diploid) by the fusion of nuclei (fertilization) and, subsequently in the life cycle, halved by the process of meiosis.

In many lineages, far more than cells come together: it is very common, for example, for entire chromosomes (one from each parent) to line up and exchange parts in the process known as *crossing over*. On the organismal level, of course, warm bodies, packets of sperm, clutches of eggs, bunches of pollen—that is, entire organisms or their parts—enjoy a fusion of some sort. Whatever the evolutionary history of sexual coming together entails, it must account for the extraordinary behavior of organisms, of their component cells, of the component nuclei of these cells, and of the chromosomes inside these nuclei. The reckoning, furthermore, cannot be attributed to any tautological need for "coming together to have sex," for, as we have seen, autopoiesis and reproduction function without a hitch in many cases in which there is no sex whatever. If there is, then, no *a priori* need in members of the same species for sexual merging and if evolution, at least at the level of microbes, never plans in advance, how and why did sexual processes come about?

# 4 · THREATS TO DNA AND THE EMERGENCE OF SEXUALITY

## Ultraviolet Light, Chemical Death, and DNA Repair

### TIME AND THE EARLY EARTH

Our narrative history of sexuality from now on will follow, insofar as possible, the chronology of life on Earth. Cosmologists, nuclear physicists, astronomers, and space scientists have colossally changed many of the most basic human beliefs. Working independently, they have produced myriads of diverse data for investigative minds to sort out and integrate. We slowly build a picture of the timescape (Calder, 1983). In the most prevalent model, the universe, forming in the biggest bang imaginable, came into being in about three minutes 13,500 million years ago (Weinberg, 1977), and it has been expanding ever since. In about a second after it was born, matter from that bang had traveled outward three light-years. Three minutes later this matter—particles but not yet atoms—heated to a billion degrees centigrade, had covered some forty light-years.

The universe is primarily made up of hydrogen and helium, the first two elements of the periodic table. The heavier elements making up the Sun and planets came later. Most were produced in the rare celestial events called "supernovae." Supernovae are extremely bright but short-lived explosions that occur as stars or stellar clouds violently collapse into neutron stars and release most of their mass in the form of energy. Atoms such as those that eventually found their way into the bodies of autopoietic beings began as products of these spectacular explosions. Such matter—spinning, expanding, exploding—became localized in many parts of the universe, including our solar system. Some cosmologists feel that not only can the origin of the Sun and its planets now be explained, but a chronology of the life history of the solar system, starting just this side of 5,000 million years ago, can be drawn up.

38

Some 4,650 million years ago a presolar cloud of gas and dust hovered in our region of the universe. Shock waves from the spiral arms of passing galaxies swept through the cloud that was to become the Sun and created a huge new star. The destiny of stars depends on their total mass: this ill-fated star giant did not last, but it provided matter for future bodies. Some 4,550 million years ago a spiral arm passed by again and made still another star. The shock wave from this supernova, the cosmogonists tell us, was enough to induce the precipitate collapse of dust and gas that converted the presolar cloud into our star, the Sun. The same series of events that made the Sun created its entourage of planets, including Earth. The Moon was apparently captured by the same gravitational forces that accreted Earth into a solid body some 4,500 million years ago. About 50 million years after that (4,500–4,450 million years ago) the concentric levels of Earth differentiated: molten iron and nickel entered the core, leaving toward the exterior the lighter elements that make up the surface crust. Energy was provided to the mix from continuous cosmic bombardments and radioactive decay from constituent elements. Well differentiated and fully made, Earth and Moon settled down about 4,450 million years ago. The meteoric and planetoid bombardment did not stop abruptly, however. We know from the cratered terrain of the Moon and inner planets that, for several thousand years at least, though probably with lessening intensity, the surface of Earth was battered and bruised by extraterrestrial impacts from meteorites of many sizes composed of a variety of cosmic materials.

The *Explorer 10* satellite has recently returned data about the output of radiation from Sun-like stars, permitting us to reconstruct the probable output of light energy of the early Sun (Canuto et al., 1982). Output was so great that, if it had remained unattenuated by the atmosphere (an atmosphere with a composition that seems most reasonable to astronomers and geologists for the early Earth), life should not have evolved at all. Deadly ultraviolet light, including those particular wavelengths absorbed by DNA, RNA, and proteins, shone down mercilessly. If, as is probable, the early atmosphere was composed of nitrogen, water vapor, and carbon dioxide, life was challenged from the time of its origin with the "danger/opportunity" of dealing with the large ultraviolet fluxes. As was often to happen later in the evolutionary story, under different circumstances, life-threatening danger became transformed into life-producing opportunity. Microbes had to deal with the large fluxes of ultraviolet radiation that pierced the early atmosphere.

Today, free oxygen forms ozone and ozone very effectively shields us from ultraviolet light of these threatening wavelengths (fig. 9). But planetary oxygen, scientists now agree, has emanated as a waste product from later life (Cloud, 1983). Most recent estimates of the maximum amount of oxygen that

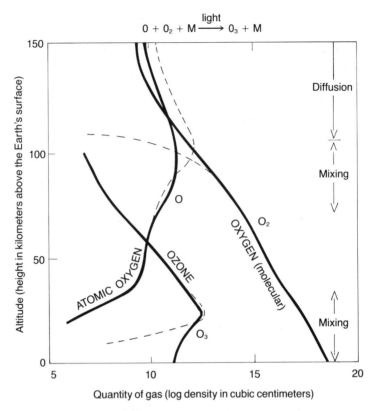

**Fig. 9.** Ozone in the Earth's atmosphere. The photochemical reaction producing ozone from oxygen is shown at the top; M = any molecule entering the reaction. Solid lines are actual quantities of the gases, dotted lines indicate calculated values based on the assumption of photochemical equilibrium. Mixing and diffusion at the altitudes in the atmosphere indicated at the right cause departure of the values from equilibrium. For details see Goody and Walker, 1972, and Walker, 1986. (Drawing by Christie Lyons.)

could have been around in the atmosphere prior to the onset of oxygen production by life are minuscule indeed, about one part in 100 million by volume (Canuto et al., 1982). Thus from the beginning of life ultraviolet and visible light threatened the very integrity of life, as will be explained below.

## FIRST LIFE AND MULTICELLULARITY

The scientific community, as well as members of the public, have been astonished during the last two decades to learn the probable age of life on this planet. We now realize that there is direct evidence that life has existed nearly as long as there has been a planet Earth. The oldest dated rocks, rocks that have never remelted since their formation, are about 3,900 million years old. They may contain evidence for life on Earth at that time—according to optimists. Unfortunately, these oldest rocks, of the Archean Eon (see table 3) have been too much heated and pressurized for fossils of living beings to have survived. They do, however, contain large quantities of carbonaceous matter. In such rocks from Greenland and Labrador, carbon in the form of graphite, with tiny quantities of more complex organic compounds, may represent traces of the early events that led to life on this planet. Yet the evidence here is hardly conclusive.

A little closer to the present, though, the evidence for early life becomes impressive. That bacteria of some kind inhabited Earth as far back as 3,500 million years ago is supported by a rather rich series of observations, growing richer all the time. From both the Warrawoona Formation, at North Pole in western Australia, and the Fig Tree Formation of sedimentary rocks, in the Swaziland System from southern Africa, a concatenation of evidence convinces us of the antiquity of life. In the late 1960s and early 1970s Elso Barghoorn, professor of biology and geology at Harvard University, and his colleagues came forth with impressive evidence (fig. 10) that pushed back the fossil record about a billion years (Barghoorn, 1971).

Few now doubt that these fossils were authentic remains of living organisms or that the early life Barghoorn discovered was bacterial. From the beginning it was clear that spheroidal unicells were present, some apparently in the process of division. Later filamentous (and therefore multicellular) bacteria were discovered both in the same material and in rocks of comparable age from Warrawoona, Australia (fig. 11). It is not surprising that multicellular life follows on the heels of unicellular forms, because offspring cells so often fail to separate from parent cells after division. Strings of cells (called "pseudofilaments" because they have no cell connections) form first, then true filaments evolve. Probably because of this failure of quick separation after reproduction, the first and simplest sort of multicellularity arose many times in many bacterial lineages (fig. 12). Some of the different multicellular forms in bacteria are shown in figure 13. Multicellularity, a characteristic of many species, is another aspect of life that is not restricted to sexual organisms, to animals and plants, or even to eukaryotes. Indeed, the term *multicellular being* does not necessarily designate an animal or plant (Margulis, Mehos, and Kaveski, 1983). Furthermore, multicellularity in bacteria may be accompanied with some differentiation in the sense that the cells of the multicellular organisms differ from each other both morphologically and physiologically (table 4).

The story of the origin of life on an early, turbulent Earth is being told in growing richness of detail (Day, 1984; Miller and Orgel, 1974). The question we are asking is how sexuality began within the earliest life. How, for example, did the Archean ultraviolet light flux affect sexuality?

**Table 3.** Geological Time*

| | EON | ERA | PERIOD | EPOCH | AGE IN MILLIONS OF YEARS |
|---|---|---|---|---|---|
| PRE-PHANEROZOIC | HADEAN | | | (Origin of Earth) | 4,500 |
| | ARCHEAN | | | (Oldest rocks) | 3,900 |
| | | | | | 2,600 |
| | PROTEROZOIC | Aphebian | | | 2,000 |
| | | Riphean | | | |
| | | | | | 1,000 |
| | | Vendian | | | 580 |
| | | | CAMBRIAN | | 500 |
| | | | ORDOVICIAN | | 440 |
| | | | SILURIAN | | 400 |
| | | Paleozoic | DEVONIAN | | 345 |
| | | | CARBONIFEROUS | | 290 |
| | | | PERMIAN | | 245 |
| | | | TRIASSIC | | 195 |
| PHANEROZOIC | | Mesozoic | JURASSIC | | 138 |
| | | | CRETACEOUS | | 66 |
| | | | PALEOGENE | Paleocene | 54 |
| | | | | Eocene | 38 |
| | | | TERTIARY | Oligocene | 26 |
| | | Cenozoic | | Miocene | 7 |
| | | | NEOGENE | Pliocene | 2 |
| | | | | Pleistocene | 0.01 |
| | | | QUATERNARY | Recent | 0 |

*Not to scale

**Table 4.** Bacterial Differentiation

| TYPE OF ORGANISM | EXAMPLES | COMMENTS |
|---|---|---|
| Endospores | *Clostridium, Arthromitus, Bacillus* | Extensive developmental cycle showing formation of new membranes |
| Prosthecae (stalk formation) | *Caulobacter* and other prosthecate bacteria | May alternate with flagellated stage |
| Heterocysts | Cyanobacteria | Nitrogen fixation |
| Thallus | Cyanobacteria, purple photosynthetic bacteria, zoogloeas | Differentiated basal cells; cubes may be formed or flat sheets of cells may cover extensive areas |
| Exospores | Actinobacteria | Borne on tips of trichomes |
| Colonial structures | *Arthrobacter*, some pseudomonads and bacilli | Complex fruiting structures in myxobacteria group, exudates from cells |
| Akinetes | Cyanobacteria | Thick-walled dissemination structures |

## THE DANGEROUS OPPORTUNITY OF ULTRAVIOLET LIGHT

Sunlight, unattenuated by ozone, posed an incessant threat to DNA integrity. Ultraviolet light at first was not separated from the visible light that powered early living systems. Yet within a geologically short time, perhaps a few hundred million years, all life on Earth became dependent, as it still is today, on photosynthesis. Staying in the sunlight was, and still is, obligate for photosynthetic bacteria, and hence for all early life. To absorb the necessary and less damaging visible light and avoid the life-threatening ultraviolet light was a problem facing all light-requiring organisms before the formation of the ozone shield.

The conservatism of living systems permits us to reconstruct their past. The first clue to the origin of sexual systems came from an observation that is well known but almost never discussed in the evolutionary literature: mutations leading to the loss of sexual ability in the colon bacteria simultaneously render these bacteria extremely sensitive to ultraviolet light. Likewise, mutations leading to the loss of the ability to cope with ultraviolet light often

44

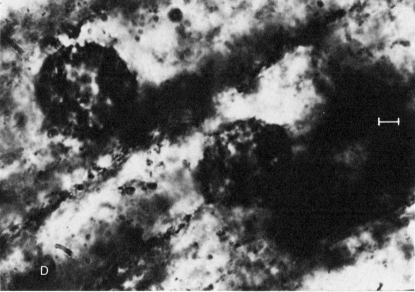

Fig. 10. Archean rocks and their modern counterparts. A. Sample of a laminated rock, a chert, from the Swaziland System that may represent the fossil remains of a stratified microbial community (actual size). B. Laminated sediment from Baja California, Mexico, known to be a stratified microbial community living and growing from the early 1960s or earlier until 1979, when it was buried by a flood (actual size). C. Archean rocks outcropping in the Barberton Mountains of South Africa (the Swaziland System). D. Microfossils like these Swaziland spheroids found in the chert rocks of the Barberton Mountains are thought to represent some of the earliest bacteria to have lived (bar = 1.0 micrometers).

45

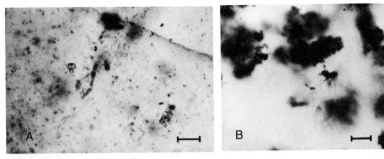

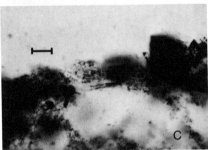

**Fig. 11.** Archean microfossils. The structures pictured here are interpreted as the remains of the oldest filamentous bacteria on Earth, approximately 3400 million years old (bar = 5 micrometers). (Courtesy of Stanley R. Awramik, University of California, Santa Barbara.)

Pseudofilament
Cells fail to separate after division, but there are no connections between cells.

True filament
Cells fail to separate after division, and connections are retained between cells.

Differentiation
Cells from same parent are specialized for different functions.

**Fig. 12.** Origins of multicellularity. (Drawing by Laszlo Meszoly.)

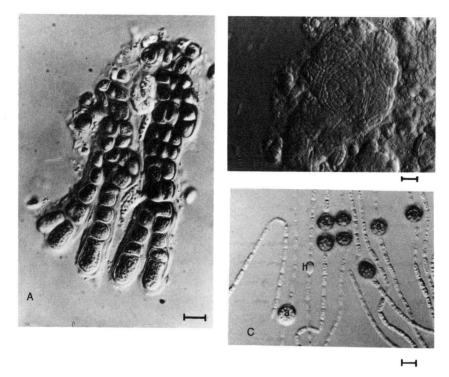

**Fig. 13.** Multicellular forms in bacteria (bar = 5 micrometers). A. *Hyella* sp., an endolithic cyanobacterium. (Courtesy of Thérèse Le Campíon.) B. Unidentified; most likely a heterotrophic bacterium. (Courtesy of Stuart Brown.) C. *Anabaena* sp., a multicellular differentiated cyanobacterium. a = akinete; h = heterocyst. (Courtesy of Stuart Brown.)

simultaneously lead to the total destruction of the genetic recombination system. *Escherichia coli* bacteria no longer able to undergo recombination are called "rec minus" mutants. These mutants, in many cases, are hundreds of times more prone to death by ultraviolet light.

A few of these cases are understood in detail. The major effect of ultraviolet light on DNA is the formation of lethal knots in the linear sequence where two thymine bases are found next to each other. Four-membered carbon rings, called "thymine dimers," are produced—as if in the steps of the ladder the bases had twisted and attached to their neighbors above and below rather than to those on the opposite side, as they should. Thymine dimers (fig. 14) knot up DNA so badly that replication and information transfer to RNA are interfered with and death ensues.

The inevitability of ultraviolet-induced formation of thymine dimers and

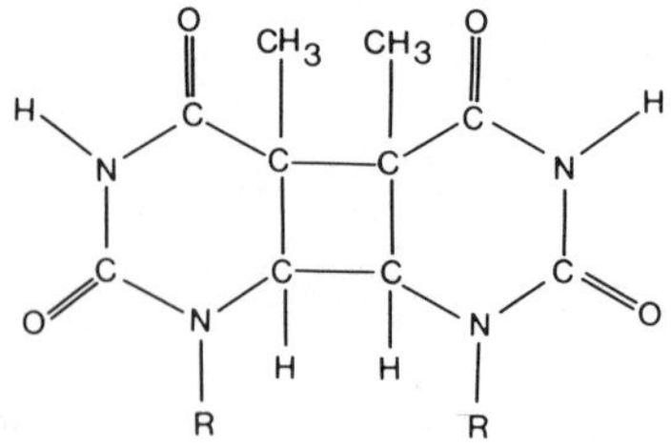

**Fig. 14.** Thymine dimers. R = radical. (Drawing by Steven Alexander.)

subsequent death must have been averted early on in the history of life on this planet. Even today there are elaborate and complex methods of enzymatically overcoming the ultraviolet death threat. These methods, applying only to the colon bacterium, in which the repair system has been best studied, are shown in table 5. A remarkable variety of mechanisms ensuring the retention of the integrity of DNA molecules exists in cells. Three classes of these have been recognized: (1) fidelity mechanisms associated with DNA replication (for example, biochemistry involving proofreading and mismatch correction processes), (2) detoxification mechanisms (primarily against oxygen), and (3) many kinds of repair (Haynes, 1985).

In retrospect, the arguments for ultraviolet radiation leading to the first sex among bacteria is simple. DNA, RNA, and protein strongly absorb ultraviolet light. The early atmosphere of Earth was composed primarily of water vapor, nitrogen, and carbon dioxide, none of which absorbs ultraviolet light in the part of the spectrum that organic macromolecules absorb (260–280 nanometers). Thus all microbes living in the open before the appearance of the ozone layer (which absorbs and therefore protects against ultraviolet light) must have been subjected to threats to the integrity of their macromolecules.

Particularly threatened were photosynthetic bacteria that required visible light. An entire battery of mechanisms for ultraviolet protection evolved, including enzymes that directly repair damage to DNA integrity. At least some of these enzymes act by removing damaged DNA sequences (such as thymine dimers) and resynthesizing DNA, using available intact DNA as a complement.

Insofar as the damaged DNA uses its complement to guide enzymatic repair, the system is merely a repair system. If no complement is available in the cell itself the DNA is irreparable and the cell dies. If DNA from any other cell source is used as a complement—and that source may be any small replicon, such as a bacteriophage, or it may be appropriate DNA in solution (called "transforming DNA")—the repair process is a two-parent one and, by definition, is a form of sexuality. The splicing and polymerase enzymes for repair became the enzymes of sexuality; DNA recombination in nature is far older than it is in the laboratory. Thus, in the evolutionary sense *ultraviolet repair preadapted bacteria to sexuality.* The source of the complementary DNA in some cases was from an entirely different cell with its own different, but homologous, DNA. When the source of the new DNA was different from that of the damaged DNA, the repair process became a form of prokaryotic sexuality: new DNA was formed from more than a single parental source.

Later, after the appearance of some atmospheric oxygen and with it the ozone screen, ultraviolet light became a less serious threat. Yet ultraviolet repair systems were still retained in many organisms because they had become part of sexual and other systems that by now served various functions in addition to ultraviolet protection. These DNA repair systems were retained for different reasons in subsequent lineages.

We conclude that the first sort of microbial sexuality was a direct response to life-threatening danger. Apparently this scenario, whereby the preadaptation to sexuality evolved as a method of survival, applies only to a subset of the members of the Monera kingdom. For example, gram-positive bacteria, including all the endospore-forming organisms (see chapter 5), seldom pair or form direct cell connections, although some gram-positive bacteria have small replicon–mediated recombination (Clewell, 1985). Though cells may come together in clumps, sexuality in these organisms is much less well understood than that in gram-negative bacteria. No gram-positive bacteria are photosynthetic. Perhaps, having no need for sunlight, they were secluded from the zones of ultraviolet radiation that applied fierce and constant selection pressure toward the refinement of DNA repair mechanisms. The atmospheric build-up of oxygen and ozone about a billion years ago was due to the spread of active, water-using photosynthesis. After the formation of an atmosphere that

**Table 5.** Cyanobacterial Mechanisms for Protection from Ultraviolet Light

Photoreactivation of ultraviolet damage
Dark enzymatic repair
Environments having high concentrations of nitrate or nitrite
Increased pigment synthesis or cell exudates that absorb ultraviolet light
Endolithic habit (penetrating and living inside rock)
Gas vacuoles and dense inclusions for regulation of depth in water column
Formation of community structures such as scums and mats

shielded the planet against ultraviolet radiation, the inducement to evolve such mechanisms must have been lessened considerably.

## THE IMPERATIVE OF LIGHT AND THE DOUBLE JEOPARDY OF PHOTOSYNTHESIZERS

Some form of resistance to ultraviolet light in the early Archean Eon was imperative and many methods that could have been used are known. The idea that resistance to ultraviolet light is an ancient legacy is supported by the observation that obligate anaerobes, organisms such as *Clostridium*, are poisoned by oxygen. The fact that organisms tend to retain their most important attributes suggests an unbroken lineage, a continuity between ancient and modern forms. The ancestors of *Clostridium* must have evolved prior to the entry of oxygen into the atmosphere, for they are far more resistant to ultraviolet light than are aerobes. Furthermore, when microbes that can be grown under either anaerobic or aerobic conditions are tested for resistance to ultraviolet radiation delivered anaerobically, the same organism is more resistant if grown afterward anaerobically than if grown in the presence of oxygen (Rambler, 1980; Rambler and Margulis, 1980). One assumes that resistance of anaerobes to ultraviolet light is well developed because these organisms evolved when such radiation was a major threat to their existence. Resistance may be maintained because the entire ultraviolet-response system has found other uses. This, however, is not altogether clear.

Recombination, presumably, is only one of many methods anaerobic microbes first developed to counter the ultraviolet threat. Another method is spore production. Spores, bacterial structures resistant to heat and desiccation, are much more hardy in the presence of threatening radiation than are the corresponding growing forms, or "vegetative structures," of the spore-forming bacteria. Spores can delay germination until sundown or until they are covered with protective layers of water, other bacteria, or organic scums. Another simple solution to the ultraviolet threat is the "sunglasses" ploy. Staying immersed in water-soluble compounds or covered by insoluble compounds that absorb ultraviolet radiation is an example of this method. Bacteria covered by the remains of other live bacteria or even bacterial debris are protected from the ravages of ultraviolet light, whereas unshielded bacteria under the same conditions die. Cyanobacteria may employ the method of living in the presence of high concentrations of nitrate or nitrite (Rambler, 1980). Some bacteria evidently protect themselves by appropriate choice of medium. Others, however, activate repair enzymes specifically for the purpose of healing the damage. At least one class of these enzymes is stimulated by ordinary visible light. If a microbe is placed in the dark after ultraviolet treatment it will die, whereas if it is placed in the light it will live. This survival in light is the result of photoreactivation, that is, repair of DNA by light-dependent enzymes. Since in nature visible light accompanies the ultraviolet, the bacteria must have availed themselves of photoreactivity from the very beginning.

Cyanobacteria are oxygen-producing, gram-negative, photosynthetic bacteria. Widely distributed and tenacious microbes, they have apparently such a large bag of tricks to protect themselves against the constant annoyance of ultraviolet light (see table 6) that they—or at least the well-studied laboratory weeds—never developed sex at all. Their relatives, photosynthetic bacteria that do not produce oxygen (such as *Rhodopseudomonas*), apparently do engage in *E. coli*–style sexuality (Marrs, 1983).

## ULTRAVIOLET-INDUCED VIRAL DISSEMINATION

The relationship between prokaryotic sexuality and ultraviolet light has been known and used as a practical tool since D'Herelle's discovery of "filterable agents," more commonly known as viruses, early in this century (D'Herelle, 1926). Although the fact is usually not mentioned in an evolutionary context, it is well known that mild ultraviolet-light treatment, such as the placement of an appropriate culture of bacteria beneath a germicidal mercury lamp ultraviolet source for less than a minute, induces the emergence from the cells of various kinds of small genetic entities. Death of the bacterial cell accompanies the release of entities called "bacteriophages." The ultraviolet-treated cell lyses, that is, bursts open. The best-studied example of this release is that of the lysogenic bacteriophages. These viruses "live" inside their bacterial hosts. Integrated within the hosts' genetic material, they reproduce whenever the host does.

**Table 6.** Bacterial Ultraviolet Repair Systems

Photoreactivation of ultraviolet damage by visible light
Dark repair: delay of growth until enzymes restore thymines from thymine dimers
Excision of damaged bases: excision repair, damaged bases spliced out
Recombinational repair: resynthesis of DNA molecule from undamaged fragments as template
SOS repair: DNA chain growth across damaged segments (an error-prone process)

*Note:* All but the first, which requires exposure to visible light, are dark-repair processes.

Under the proper conditions the latent phages become active and destroy their hosts. The entire activity, known as "lysogeny," or "phage burst," is routinely induced by placing the lysogenic host cells beneath an ultraviolet lamp. If we regard viruses as part of a legacy of ultraviolet-induced cell destruction, it should come as no surprise that even today ultraviolet irradiation leads to the emergence of bits of genome. Inside a phage particle (virion) the dormant DNA is safe within a protein case. It is reasonable to assume, then, that the phage particle, often with a few bacterial genes strung onto it, has a better chance of ultimately reaching a safer realm than do naked bits of genome exposed directly to irradiation.

Not only bacterial viruses but also animal and plant viruses, as well as various plasmids (which may be thought of as viruses without the usual protein coats), will often burst out of cells that are exposed to ultraviolet light. Dense genetic particles with protein coats, carrying a portion of the genome of the autopoietic host, are protected from thymine-dimer production. Safe from lethal light, the genes are available for eventual penetration and integration into the linear order of a second autopoietic host.


## INTRINSIC SEXUALITY OF VIRUSES

Often viruses and other small genetic entities carry with them bits of informational DNA that they have picked up from their former hosts. This DNA may code for one or another useful trait. Not being autopoietic, these entities do not survive unless they enter an autopoietic host. When they infect such new hosts, as is well known, they may combine their DNA with that of their new host. More specifically, the DNA of the viruses and plasmids, by the action of a battery of appropriate enzymes, becomes integrated into the linear genetic order of the chromoneme of the bacterial cell. The release by one bacterium and subsequent uptake of viruses by another probably evolved as a response to ultraviolet light and to other dire environmental conditions as well. Since the fundamental process of viral transfer between hosts also involves the formation of a new piece of DNA from more than a single parental source, it is like DNA repair of ultraviolet damage, a form of chromonemal sexuality.

Sexuality of viruses has been known as long as viruses themselves have been studied. A clue to this intrinsic sexuality was observed upon mixing a virus of one sort (one, say, that causes small, ruffled plaques in its bacterial host when it bursts out) with that of another (one causing large, smooth plaques upon bursting). The mixture of viruses produced not only copies of the parental types but also "recombinant" types (ruffled, large types and smooth, small ones). The reason for the recombination is the tendency of viral DNA inside a host to recombine in such a way that the viruses coming out carry DNA from different sources. The rules governing the proportions of parental and recombinant forms of viruses are neither universal nor simple. They can change with conditions that affect the chemistry of DNA. Suffice it to say that if the conditions for subsequent life are permissive (that is, if all recombinants—smooth, rough, large, and small—can survive) recombinants *do* survive. Sex, in the sense of viral recombination, was not necessary for reproduction or anything else; it simply occurred in the course of survival. The enzymes to break, patch, repair, and recombine DNA were present and functioning.

We have indicated that prokaryotic sexuality arose as a response to the threat of ultraviolet-forced disintegration of nucleic acids. It is likely that an analogous tale can be told of responses to chemical threats as well. These include threats of toxic concentrations of mercury or manganese, the resistance to which is known to be borne on the sort of small replicons called "plasmids." But, because the detailed organic and metallic chemistry of the early Archean Eon is more difficult to reconstruct than is radiation quality and flux, we have confined ourselves to the example of ultraviolet light.

In summary, then, prokaryotic sexuality—recombination on the DNA level—is best understood as a possible response to ultraviolet irradiation and other threats to DNA survival. These survival mechanisms involved the borrowing of an undamaged DNA. To be usable as a complement from which a good DNA copy could then be made, the second DNA had to be recognizably similar to the first. In this step alone we see the appearance of at least a second parent.

Prokaryotic sexuality, first an enzymatic response to ultraviolet or chemical threats to the linear integrity of DNA, later became co-opted for other tasks. An example of such later co-opting can still be seen today. The distribution throughout the environment of certain types of plasmids and viruses is revealing. It has been observed (Silver, 1983; Rosson and Nealson, 1982a, 1982b; Lidstrom, Engebrecht, and Nealson, 1983) that, in environments bearing toxic quantities of metallic compounds and organic poisons (including antibiotics), the genetic factors carrying resistance to these insults are borne on viruses, plasmids, or other small, mobile replicons. These entities quickly reproduce and are passed from vulnerable cell to cell in direct response to the environmental toxin. (For a review of genetic transfer by such transient entities in prokaryotes, see Sonea and Panisset, 1983.) Recombination of microbial DNA, a legacy of response to traumatic ultraviolet radiation in the Archean Eon, is the first step in the long and winding sexual pageant of planetary life.

# 5 · RECOMBINATION AND BACTERIAL MATING

## Worldwide Evolutionary Changes through Microbial Sex

CONJUGANTS

Whether most bacteria are capable of mating is not known; neither is the extent of bacterial conjugation in nature (Sonea and Panisset, 1983). Some bacteria mate with enough consistency, however, that the process can be studied in the laboratory. In bacterial mating, or bacterial conjugation, two cells must first come into physical contact so that their outer membranes touch (fig. 15; see also fig. 7).

The mating is always polarized, never reciprocal: one mate donates genes to the other mate. The donor is conceptually the "male," because the genes travel from "him" to be received by "his" partner, conceptually a "female." There are no exchanges in bacterial conjugation, even or uneven: in each mating a donor must conjugate with a recipient. Immediately after the mating a recipient, having received a small fragment of DNA that confers "maleness," may be converted into a donor. Thus the genes for gender themselves are sexually traded.

Biologically speaking, *maleness* refers to gender, functional and correlated morphological differences that distinguish mating partners. We shall use the term *gender* in this sense. Yet, although the donor in a bacterial mating can be considered to be of the male gender, there are no visible differences between donor and recipient prior to the mating event. Furthermore, since a donor can quickly be converted to a recipient, and a recipient to a donor, gender differences in bacteria are unstable and difficult to detect. This genetic determination of gender depends, at least in *E. coli*, on the physical acquisition of a single small, viruslike replicon called the *F-factor* (the *fertility factor*). Bacteria bearing this particle, which is little more than protein-coated DNA at most several genes long, are donors, whereas those lacking it are recipients. The recipients have little hairs, pili, on their surfaces that permit them to take up F-factors. Immediately upon receipt of the particle a recipient is converted into a donor. For this reason the strongly charged labels *female* and *male* tend to be replaced in the bacterial literature with the more bland, technical terms *donors* and *recipients*. Donors become recipients simply by losing the F-factor. Many other genetic entities besides F-factors—such as prophages and plasmids, all small replicons—are traded between bacteria. Indeed, the extreme promiscuity of gene transfer in bacteria renders the idea of fixed sexes meaningless.

Unlike the more familiar situation in meiotic sex, in which parents contribute their genes equally to make a new offspring, there is no equal contribution of genes at all in a bacterial conjugation. The recipient bacterium may receive one, nearly all, or even all, of its donor's genes. Because the distribution from very few to very many, depending on the conditions of the mating, is so broad, almost never do half the genes in a given mating come from the recipient and half from the donor. No reproduction at all is involved, only genetic change. The donor rounds up his mate and forces "his" genes into "her." Under many

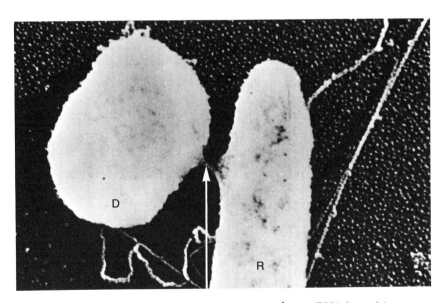

**Fig. 15.** Conjugating bacteria. R = recipient; D = donor. DNA (arrow) is presumably being passed from donor to recipient.

conditions neither the donor (because "he" lacks genes present only in the recipient) nor the recipient (because "she" lacks genes present only in the donor) survives. The fact that the only survivors are recombinant bacteria, containing the genes from two or more parents that are now needed to cope with the new changed environment, ensures natural selection of the recombinant bacteria and thus persistence of the donation process.

## GENE DONATION

Several factors are known to influence the number of genes actually donated in a mating. When other factors are equal, more genes will be transferred to the recipient the longer mating is permitted to proceed (an hour instead of three minutes, for example). Matings performed at higher (but not too high) temperatures will lead to the transfer of more genes than do those at lower temperatures. Chemical factors, such as salt concentration, can inhibit the number of genes transferred.

Bacterial conjugations never involve cell fusions. They are DNA recombination events in the sense that only some genes, small replicons or part of the genophore (the chromoneme), are passed. Recipient bacteria incorporate the donor genes they receive into the linear order of their own DNA in the same way that they incorporate viruses, plasmids, and other small replicons. Enzymes such as nucleases, nucleotide polymerases, ligases, and so forth are employed in the recombination.

A bacterial donor that transfers all its genes to a recipient will of course die from lack of genes. Indeed, if a donor does not keep a copy of a complete genome for itself it will not survive the mating. Another aspect of this "unidirectional" gene transfer is that the recipient bacterium will at first have not only a copy of its own genome but second copies of genes received from the donor. In the recombinatory events that follow mating the recipient's genes must be sorted out so that it ends up with only one copy of each gene per genome, regardless of origin. To achieve this end, donor genes are incorporated in proper order in the genome and duplicate genes are digested away by nuclease enzymes. The precision of this process is remarkable and must involve recognition of DNA homologies and other chemical details about which we are quite naive.

The presence of the fertility factor determines the ability to transfer genes and thus is entirely correlated with the donor state. In normal cases the F-factor replicates faster than the rest of the bacterial genophore. It thus spreads, viruslike, throughout the population of bacteria. For this reason sex change is rampant: large numbers of recipients may quickly be converted into donors by the simple acquisition of a benign "viral infection." Treatment of donor cells with acridines, nitrogen-containing organic bases that can interca-late (squeeze between) the normal bases of the DNA molecule, can produce aberrant F-factors. Because these replicate more slowly than the rest of the genome, F-factors can be lost with such high frequency that almost an entire bacterial population is converted to recipients. Loss of the F-factor leads directly to sex change, turning genetic "females" into "males." Heat, too, may be responsible for gender change in bacteria.

Microbial geneticists determine sexuality in bacteria by experiments in natural selection. The strategy, in general, is to grow a clone of bacteria sensitive to an antibiotic or to heat or distinguishable in some other way. Examples include bacteria unable to grow unless an amino acid or some vitamin is added to their medium. A second, different bacterial population is then cloned—say, bacteria that are resistant to sodium azide, an inhibitor of oxygen respiration. The bacteria of the different kinds are mixed and recombi-nants are selected. In the simultaneous presence of sodium azide and the absence of the required vitamin, neither parental type of bacteria can survive. Only recombinants containing genes from both parents live: those with the gene conferring the metabolic capability to make the vitamin and the gene conferring resistance to sodium azide. This sort of experiment, which permits only recombinant bacteria to live, has been repeated thousands of times by hundreds of investigators.

It seems very clear that strong selection pressures—starvation for want of a vitamin, the presence of toxins, and so forth—will permit the survival of recombinant bacteria (and the recombination process itself) and simul-taneously select against parental forms. We have seen in rough outline in the previous chapter how the fundamental set of enzymes that repaired ultra-violet damage evolved. These are the same enzymes that permit recombinant DNA to form inside cells. Once these enzymes had ensured survival of ultra-violet-damaged DNA, the same enzymatic machinery was utilized for bacte-rial conjugation under circumstances in which only recombinants survived. The steps from incorporation of small genetic entities to incorporation of DNA from mating partners are few. The major innovation is the ability of bacteria with F-factors to produce some surface substance making them attractive to organisms lacking the F-factor. Only a small number of genes was required for this new step.

## DNA REPAIR AND SEX AS RECOMBINATION

Autopoiesis absolutely depends on the integrity and continuity of the DNA molecules inside cells. In the past two decades molecular biologists have continually been surprised and impressed at the number, complexity, and diversity of mechanisms that cells possess to repair injuries to their DNA (Kornberg, 1980). Not only is damage from ultraviolet light handled with alacrity and high priority, but a whole slew of DNA repair mechanisms exist to deal with other kinds of damage. For example, chemically damaged bases in DNA are replaced in a complicated, multistep but accurate process called *excision repair* involving "cut-patch-cut-seal" tactics. Damaged DNA, which can be recognized by the distortions in the regularity of the double helix, is removed by the action of an endonuclease, an enzyme that cuts nucleotides from one of the complementary helices inside, not at the ends of, DNA. A new strand of about twenty nucleotides is then synthesized by a polymerase enzyme, pushing the old twenty nucleotides out of the way. This excised fragment is degraded by other enzymes. Finally, the new strand is joined to the intact DNA by a ligase enzyme. The incision activity, at least in the colon bacterium, involves the products of at least three different genes called "recA," "recB," and "recC." They are named for the observation that mutations in any of these genes render the cells unable to recombine; simultaneously they become sensitive to ultraviolet light. Excision repair can be considered an Archean legacy of response to ultraviolet threats.

That bacteria are richly endowed with the ability to repair ultraviolet-damaged DNA is directly related to the origin of sex. In case the information in the original DNA helix is lost by damage there are only two possibilities: death, or the use of a second, complementary DNA helix to supply the missing information. Originally the missing information may have been provided by small replicons that broke away from parental DNA as ultraviolet light impinged upon the bacteria. At first the back-up genetic information probably came from extra copies in the same cell. When, however, a helix from a second individual was used, new DNA derived from multiple sources was formed. This modification of the repair process became bacterial sex. Excision repair is a system that can be used if only one of the two helices of the damaged DNA is destroyed; the second helix must be intact. That is, since DNA is a complementary molecule, the newly synthesized DNA uses the undamaged helix as a source of information. In evolutionary terms, the appearance of excision-enzyme mechanisms to repair damaged DNA is a preadaptation to bacterial sex in which an entirely different DNA molecule is used for the source of the information to repair the damage. In general, replication is totally halted by damage in DNA. Either the damage is repaired or the cell dies.

There are other systems besides excision repair that result in the healing of only a single DNA molecule. For example, a system called "error-prone" or "SOS repair" allows DNA replication to carry on even in a damaged molecule.

The DNA polymerase molecule involved in this process apparently permits DNA with mispaired bases in it to replicate. Of course the consequence of this is the production of many mutants. The SOS repair system, with enzymes that tolerate errors, is apparently induced as a result of ultraviolet damage to DNA. After ultraviolet irradiation the cell synthesizes the machinery it needs to repair the damage sloppily. What is very interesting to our narrative is that, in order to succeed even in sloppy repair, the gene recA, known to be absolutely required for bacterial mating, is also needed. The gene product of the recA gene is a protein having several simultaneous functions, all related, we believe, to its historical function of DNA repair. When treated with ultraviolet light, bacteria produce up to tens of thousands of copies of recA protein molecules. This protein, an ATPase, breaks the bacteriophage repressor protein into two parts, rendering it inactive. As a result phage particles, which leave the bacterium containing fragments of the bacteria's genome, are released. This same protein acts directly in the repair of x-rayed DNA and of drug-induced DNA damages and in the SOS functions (Roberts et al., 1985). Again, a complex repair system is genetically related to the sexual system— and related to tolerance of potentially lethal ultraviolet light.


EVOLUTION OF CONJUGATION

Given the continual threats to DNA integrity on a hot, upheaving Archean Earth, the selection pressure on bacteria must have been severe. From what we understand now, the base thymine in DNA strongly absorbs ultraviolet light and tends to react with any nearby thymine by making a thymine-dimer "knot" (see fig. 14), the major type of ultraviolet damage to the DNA double helix. As the replicating enzymes try to copy the DNA they come to the knot and stop. The enzymes chemically act on the nucleotide precursors but they do not become incorporated into the growing chain, so replication stops, like a car motor idling. Although the enzymes could ignore the knots and begin the replication process beyond them, the resulting new DNA strand would have large gaps in which information would be lost forever. This must have resulted in death many times, even for cells already equipped with excision and SOS repair systems.

Even today a kind of repair system involving genetic recombination has been identified in cells: a second, perfectly healthy DNA double helix is used as the source of information for the damaged piece. The appropriate segment of the healthy DNA helix from a donor is excised and inserted into the gap formed by the removal of the thymine dimers. During the next round of replication the donor's DNA is copied. This of course results in a new recom-

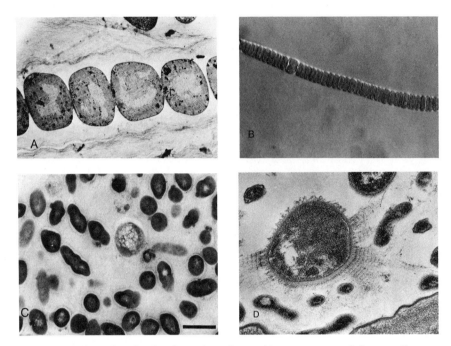

**Fig. 16.** Polysaccharide sheaths and exudates of bacteria. A. *Nodularia,* a filamentous cyanobacterium from the microbial mat at Laguna Figueroa, Baja California, Mexico, forms a conspicuous polysaccharide sheath even in laboratory culture. Cells are 2 micrometers × 6 micrometers in size; width of filaments is 6 micrometers. (Courtesy of John Stolz; see Stolz, 1984.) B. *Nodularia* with heterocysts. Width of filament is 6 micrometers. (Courtesy of John Stolz.) C. Some bacteria, like this bright red, heterotrophic, gram-negative rod, form hard, tight colonies when they grow in agar because the cells produce an exudate, seen as fuzz surrounding each cell (bar = 1 micrometer). (Courtesy of Stuart Brown; see Brown et al., 1985.) D. Bacterial striated exudate, bacterial cell diameter approximately 3 micrometers. (Courtesy of Dr. David Chase.)

binant DNA carrying genes from two parents. Furthermore, the gap left in the donor is copied and rejoined by polymerase and ligase enzymes because the complementary helix of the donor's DNA can serve to supply the genetic information.

That genetic recombination began as a part of an enormous health delivery system to ancient DNA molecules is quite evident. Once healthy recombinants were produced, they retained the ability to recombine genes from different sources. As long as selection acted on the recombinants, selection pressure would retain the mechanism of recombination as well. Recombination continued to be selected.

Even when a recombinant microbe is greatly outnumbered by parental forms, making up, say, one bacterium in a billion (a situation very common in the laboratory), the recombinant may be preferentially preserved. Since the one in a billion will become the only survivor under certain harsh conditions, there is often strong selection pressure to retain the capacity to recombine DNA. Those organisms capable of a flexible reception and genetic integration of new life-saving DNA were those which persisted. The mixing and matching of genes ensured these bacteria, apparently all gram-negative forms, of a future. Gram-negative bacteria are distinguishable in the laboratory by their lack of retention of the purple stain developed by the nineteenth-century Danish physician Hans Gram. They stain pink. Gram-positive bacteria retain the crystal violet stain and therefore stain purple. The stain reflects the absence, in gram-positive cells, of an outer lipid protein membrane around the bacterial cell. This outer membrane is probably directly involved in the cell-to-cell interaction in bacterial conjugation. It may account for the fact that conjugation has been difficult to detect in gram-positive bacteria.

Apparently some lineages of bacteria were never fortunate enough to inherit the DNA repair-recombination system. As far as we know, many groups of bacteria, especially gram-positive organisms, do not engage in sex at all. Many of these gram-positive forms produce spores and other resistant cell products, such as polysaccharide sheaths and exudates (fig. 16). Perhaps the strategy for survival of these gram-positive bacteria bypassed the mixing and matching of genes. Well equipped to resist environmental threats by forming hard coatings and other means, these organisms tend to wait. The gram-negative forms, unable to wait, frantically pass their genes. Although, as usual for biological scenarios, most organisms died, those bearing the winning combination of genes for a particular environment at a particular moment were suited to continue passing their genes on. Their descendants provide us with laboratory material for the study of bacterial sexuality today.

# 6 • THE EMERGENCE OF PROTISTS

## Symbiotic Bacteria and Organellar Sex

NEW CELLS

Present-day members of the kingdom Protoctista are as diverse as and more numerous than members of the kingdom Animalia or Plantae. Yet our knowledge of protoctists is still in its infancy. Nonetheless, a study of protoctists, and especially of the unicellular protoctists, by definition protists, is indispensable to an understanding of the origin of meiotic sex. The protists undergo more varieties of meiotic sex than any other kind of organism. All are capable of single-parent, asexual reproduction. In addition to the asexuality standard in this group, some protists have two-parent sexuality in their life cycle. Some divide mitotically just like cells of our bodies cultured in the laboratory. Other cells undergo reduction division (meiosis) and subsequent fertilization, a sexual life cycle similiar to that of most animals and plants. Yet some protists divide by processes that are clearly simpler than mitosis and meiosis, and still others die whenever they atavistically engage in sex. The protoctist world thus provides us with a reflecting mirror into which we can look when speculating on the origins of mitosis and, later, of meiotic sex. Since mitosis and meiosis undoubtedly evolved in protoctists, we have included here an introduction to the biology of each of the major groups of this diverse and important kingdom.

To understand the differences—and the similarities—between the ancient bacterial style of sexuality that we have been discussing and modern modes of sex, such as the meiotic sexuality of animals, we must first recognize the evolution of a new kind of cell. This was the eukaryotic cell, or protist. In the thirty years since the French zoologist Edouard Chatton first pointed it out, it has been confirmed that the greatest division in biology is not between animals and plants but between cells with nuclei (eukaryotes) and cells without (prokaryotes). Prokaryotes, all bacteria, include organisms such as E. coli, Leuconostoc (fermenting organisms that make yogurt from milk), Gonococcus (spherical bacteria associated with the venereal disease gonorrhea), and treponemal spirochetes (thin, motile helices found in the tissues of people suffering from syphilis). Eukaryotes, either as single cells in the form of microscopic protists, such as green algae or ciliates, or in compact populations of cells that together take the distinct shapes of cats, penguins, rosebushes, mushrooms, seaweeds (protoctists) and other familiar organisms, engage in the meiosis-fertilization form of sex that is evolutionarily very different from the "gene injections" of bacteria.

Not only are eukaryotic cells structurally more complex than bacteria, but their strategies of survival are more intricate. As opposed to prokaryotes, eukaryotic cells have the ability to engulf whole live bacteria and protists by phagocytosis ("cell eating"). They are also capable of removing protein particles from solution, taking them into narrow channels deep inside themselves (pinocytosis, or "cell drinking"). Many eukaryotic microorganisms can form cysts, thick-walled protective structures that allow the organisms to wait out periods of cold, heat, starvation, and desiccation. These cysts, called "spores" when produced by fungi or obscure funguslike protists, germinate into growing, reproducing organisms under the proper conditions. Some eukaryotic cells of protists or animals even form and rapidly deploy parts of themselves, "poison darts," trichocysts, toxicysts, or nematocysts, with which they stab their prey. These many kinds, probably not directly related, are called "extrusomes." There are clearly many intricate ploys within the province of single eukaryotic cells.


## MITOSIS

The most important procedural and morphological difference between the two great groups is the mode of cell reproduction. Prokaryotes, which contain free-floating strands of DNA comprising their nucleoids, reproduce by splitting in two (fission). The newly and continuously synthesized DNA while attached to a cell membrane is distributed to each offspring cell. In contrast, eukaryotes carry their replicative DNA, intertwined with histone proteins, in large, visible bodies, the chromosomes. Nearly all—but, importantly, *not* all—eukaryotes undergo the far more complex process of *mitosis* when they divide. This process, which involves no sex at all, is responsible for the reproduction of unicellular eukaryotes (protists, by definition). In organisms such as flowering plants and all mammals that reproduce by two-parent sex, mitosis is the means of growth.

Mitosis, called "the dance of the chromosomes" in the nineteenth century, involves the formation of visible, countable chromosomes and their movement first to a plane in the center ("equator") and then to the poles of the spherical cell (E. B. Wilson, 1925). The movement is along a fibrous, spindle-shaped structure called the *mitotic apparatus*. Seen with the electron microscope, the mitotic apparatus is composed of hundreds of microtubules, each of which is 240 angstroms in diameter and composed of tubulin proteins, about which we will have more to say later. Mitosis (mitotic cell division) is always the process by which plant, animal, and fungal cells make more of themselves during growth. It can also be regarded as a major mechanism of self-maintenance in multicellular beings: in association with cell metabolism, mitosis accounts for autopoiesis on the organismal level. Organisms not only grow but maintain themselves (for example, in the reproduction of skin cells) by mitotic division. Meiotic sex in species of organisms that regularly have two parents is, at the cellular level, clearly an evolutionary derivative of mitosis. The relation of mitotic cell division, which involves an equal distribution of genes to each offspring cell, to meiotic cell division, which involves a halving of the set of genes, is analogous to that between the expansion of a village into one about twice the size ("mitosis") and subsequent colonization to form a new village by the deployment of about half the inhabitants of the first village to a new location ("meiosis"). The production of a human infant by sexual means is still primarily a form of cellular expansion because growth via many mitotic cell divisions has occurred.

Human growth events involve production of sperm and eggs, fertilization (conception), fetal development, birth, child development, adolescence, and maturation. Mitosis occurs throughout all the processes, whereas meiosis is involved only in two cell divisions prior to formation of eggs in females and sperm in males. The cell-level products of these activities are clearly distinguishable.

Eukaryotic organisms from amoebae to elephants, by definition composed of cells with membrane-bounded nuclei, are considered, because of a long list of common traits, to be derived from a common eukaryotic ancestor. The cells of water ferns, ginkgo trees, and men, for example, contain similar organelles (distinctive subcellular structures): mitochondria, chromosomes, and Golgi bodies. (Golgi bodies are membranous structures called "dictyosomes" in plant cells and some protist cells.) The ultrastructure of the sperm tails of

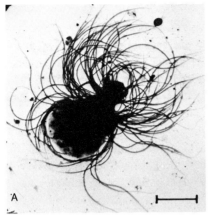

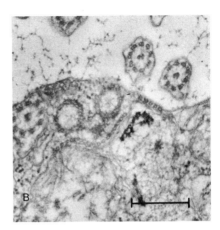

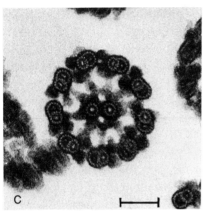

**Fig. 17.** Examples of sperm tails. A. Fern plant *Marsilea* sperm cell is shown in this negative stain electron micrograph. Bar = 10 micrometers. (Courtesy of Ikuko Mizukami and Joseph Gall.) B. Ultrastructure of sperm tail (undulipodium in fern plant). Bar = 0.5 micrometers. (Courtesy of Ikuko Mizukami and Joseph Gall.) C. Animal sperm tail (undulipodium). Bar = 50 nanometers. (Courtesy of R. W. Linck.)

these organisms is entirely held in common. Ginkgo trees, for example, have sperm cells that grow sperm tails. They are produced in fleshy male, conelike structures that hang from branches. Water ferns such as *Marsilea* produce sperm packets and release them into the surrounding lake water. So do sea urchins, echinoderm animals related to starfish. The detailed structure common to sperm tails can be visualized with a transmission electron microscope, as shown in fig. 17.

## ORGANELLES

The near universality of the microtubules of mitosis, of sperm tails, of cilia, and so forth is one reason for the idea that protists, animals, fungi, and plants had a common ancestor equipped with microtubules. The differences in the origins of groups may be seen by examining other cell parts. For instance, the most conspicuous and significant difference between the cells of animals and those of plants is the presence of plastids. The leaf cells of ginkgo trees and nearly all plants contain these photosynthetic organelles, which come in several different colors, depending on the organism: for example, green chloroplasts of flowering plants, golden yellow chrysoplasts of chrysophyte algae, rhodoplasts of red algae, and phaeoplasts of brown seaweeds.

From about one-half to ten micrometers in diameter, plastids are present in all photosynthetic eukaryotic cells, whereas animal cells lack them. Nonetheless, because of the great number of common features, it is clear that people and trees—indeed all animals and plants—have common eukaryotic ancestors. The acquisition of plastids occurred in one lineage (that leading to seaweeds and ginkgo trees) and failed to occur in a second lineage (that leading to mammals) of similar eukaryotic cells. By the time these ancestral eukaryotic cells split into several major evolutionary lineages (plants, animals, fungi, and a variety of protoctists), they already contained their characteristic organelles—plastids, mitochondria, and undulipodia (wavy structures composed of a distinct pattern of nine pairs of microtubules)—structures that are never present in bacteria.

Mitochondria, present in the cells of nearly all eukaryotes, and plastids, present in the cells of all photosynthetic eukaryotes (algae and plants), are two particularly important classes of eukaryotic organelles: they generally arise by direct division from parent organelles. Even when they develop from smaller bodies (promitochondria or proplastids, respectively), the mature bacteria-sized mitochondria and plastids develop from their respective preexisting parental organelles. Interestingly, both mitochondria and plastids have been documented in the laboratory as having sexual encounters resulting in genetic recombination. Further, it is now known that DNA sequences from these organelles enter both each other and the nucleus of the cells in which they are found (Lewin, 1984). Mitochondria and plastid genetic recombination can be demonstrated at a time when the cells in which they reside reproduce by mitotic (nonsexual) division (Gillham, 1978). The implication here—and it is only one of many clues giving similar hints—is that these organelles were once free-living bacterial entities: their type of sex, the recombination of DNA molecules, is a holdover from their former independent lives.

There are a few exceptions to the generalization that mitochondria are present in all eukaryotes. Some amoebae, such as *Pelomyxa palustris*, *Entamoeba histolytica*, and certain mastigotes of termite hindguts, such as *Trichonympha*, lack mitochondria. Symbiotic bacteria are present inside some of these protists that may function in the role of mitochondria. Methanogenic bacteria, residing inside the giant host amoeba *Pelomyxa*, receive hydrogens from the host carbohydrate food molecules (Van Bruggen, Stumm, and Vogels, 1983). Presumably the methanogenic bacteria remove *Pelomyxa*'s hydrogen as waste, combining it with $CO_2$ derived from fermentation to produce methane gas. Although eukaryotic, *Pelomyxa* thrives in the absence of oxygen.

The vast majority of eukaryotes depend metabolically on their mitochondria, and these mitochondria require oxygen; thus it is common that those eukaryotes stressed with the lack of oxygen are those in which other bacteria have apparently replaced mitochondria and their function. *Metopus contortus*, or at least strains of it found in sulfide-rich muds, also lack mitochondria (Dyer, 1984). Other ciliates, such as *Sondaria*, found in anaerobic muds, or *Plagiopyla*, found in the guts of sea urchins, have conspicuous bodies in their cytoplasm that greatly resemble bacteria but cannot yet be grown in culture. Thus no one is sure that they are bacteria. *Plagiopyla* and other members of its family in particular have very conspicuous striped organelles that apparently are two types of bacteria or mitochondria alternating with bacteria (fig. 18).

Termite hindgut protists harbor bacteria on their surfaces, in their nucleus, and in their cytoplasm, although whether any of these function to replace mitochondria is not known (fig. 19). Some bacteria inside protist cells, such as the intranuclear symbionts in *Trichonympha* (found in the hindgut of the dry-wood-eating termite *Incisitermes minor*), may actually be responsible for cellulase production. Cellulases are enzymes that break down cellulose, the major fibrous component of wood. Other symbiotic bacteria of these protists, such as those associated with endoplasmic reticular (internal) membranes in *Mixotricha paradoxa*, may be mitochondria replacements.

Mitochondrial membranes are composed of several hundred respiratory enzymes. Mitochondria make use of oxygen as acceptors of the hydrogen atoms from food molecules, in much the same way as do many oxygen-respiring bacteria. Whereas prokaryotes show a great diversity in the enzymes of their respiratory pathways, mitochondria, from protists to porpoises, extract and use energy in the same way. Beginning usually with food molecules in the cytoplasm (carbohydrates like dicarboxylic acids, such as malic, succinic, and fumaric acids), carbon, oxygen, and hydrogen atoms are removed and the hydrogen reacts with oxygen so that the end products of eukaryote cell respiration are carbon dioxide and water. The energy released from food breakdown is stored in ATP nucleotide molecules, components of DNA and RNA. ATP produced along the way is used for many processes, including biosynthesis of nucleic acid and protein, active uptake of more food molecules across cell membranes, phagocytosis, undulipodial movement, and mitotic cell division. The mitochondria are the sites of ATP energy production in

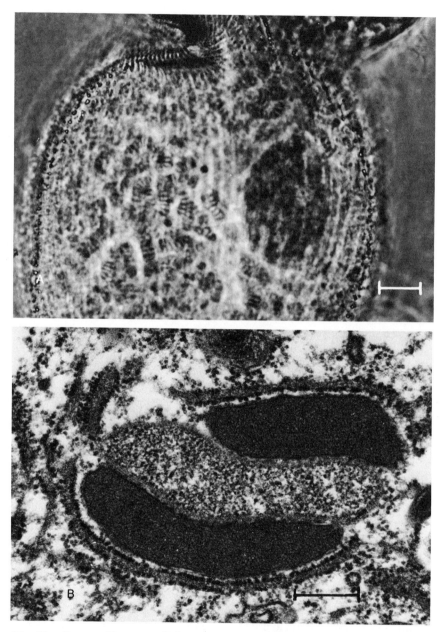

**Fig. 18.**  Bacteria-like intracellular organelles of *Plagiopyla* (a ciliate). A. Light micrograph of the ciliate showing enigmatic cytoplasmic bodies (bar = 10 micrometers). B. Electron micrograph of bacteria-like intracellular organelles of *Plagiopyla minuta* (bar = 1 micrometer). (Unpublished photo, courtesy of Jacques Berger, University of Toronto.)

nearly all eukaryotic cells; they have been called cellular "power stations." Mitochondrial membranes also provide enzymes (for example, squalene oxidase) for combination of metabolites with oxygen in the production of steroids.

The common way in which mitochondria are involved in gaining energy from food by combining with oxygen also supports the idea that all modern-day multicellular eukaryotes had common, eukaryotic, single-celled ancestors. These ancestors lived when the first free oxygen began to enter the atmosphere in massive amounts. As cyanobacteria (confusingly still referred to as "blue-green algae" in many textbooks) wafted waves of oxygen over the two-billion-year-old Earth, many heterotrophic microbes (those that do not produce their own food) developed that could efficiently utilize oxygen, a gas that was once toxic to all life and that is still poisonous to anaerobic organisms. Many eventually evolved to the point where the oxygen was required for their metabolic pathways.

There are many clues suggesting that mitochondria originated from free-living bacteria. This idea, now considered a fact by most of those interested in such matters, was expressed early in this century (Mereschkovsky, 1905; Wallin, 1927). Based on much more recent biochemical and morphological information, several models by which the intricate cellular symbioses developed have been suggested (Margulis, 1981). The large, anaerobic host bacteria that became the nucleocytoplasm could have been like *Thermoplasma*, hardy microbes of an ancient lineage that survive today in scalding, highly acid water (Searcy, Stein, and Green, 1978). The archaebacterium *Thermoplasma acidophila*, although a prokaryote, is similar to eukaryotes in that it has histonelike protein surrounding its DNA that protects it from acid degradation (Searcy and Stein, 1980). Histones, which form the core of nucleosomes that make up the chromosomes, are characteristic of eukaryotes. Perhaps ancestral *Thermoplasma* ingested but failed to digest oxygen-respiring microbes. These microbes could have become protomitochondria and eventually mitochondria. The hosts, containing symbionts, then developed into an extraordinarily successful lineage. Combining the hardiness of *Thermoplasma* with the oxygen-processing energy potential of mitochondria, the symbiotic microbes entered and established for themselves rich biological niches unavailable to other organisms.

Another model of the probable ancient alliance that took place in the ancestors to all eukaryotes is one of predator and prey. In this scenario aerobic microbes like the predatory, respiring *Bdellovibrio* invaded their prey in the devastating manner of parasites. Those prey that survived and left the most offspring were those that could best tolerate and eventually evolve a cure to

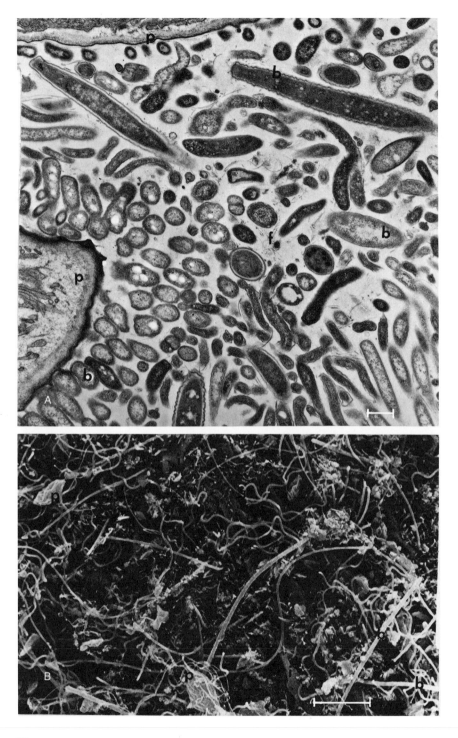

their new "disease." The pathogenic invaders, eating excess food in the prey cytoplasm, were eventually supported as a population of internal symbionts. The waste products floating in the cytoplasm of the healthy prey provided the invaders with food—a sort of unwritten invitation to remain. But for the symbionts not to die, it was necessary that they not kill their prey. Those ancestors to mitochondria that effectively used available oxygen, but at the same time kept it away from their prey's DNA, also saved themselves from destruction. Alternatively, if the prey had developed membranes to protect itself from mitochondrial oxygen, it could have saved itself from the ravages of oxygen on its nuclear DNA. This may be the origin of the omnipresent nuclear membrane, the defining feature of all eukaryotes today. Indeed, the outer membrane of mitochondria is continuous with the nuclear membrane; unlike the inner membrane, it is synthesized by the nucleocytoplasm, perhaps a latter-day version of the prey, not by the mitochondria themselves.

As oxygen levels in the atmosphere increased due to the expansion of photosynthetic microbes, such partnerships would have been favored. Contrary to its relatives in anaerobic retreat, the *Thermoplasma*-like microbe, after harboring oxygen users, could roam freely in rising concentrations of oxygen. The greater the concentration, the greater the selective pressure for more protomitochondria and membranes protective of nucleic acid. With the elapse of time, the internal enemies of the prey evolved into microbial guests, and, finally, supportive adopted relatives. Because of a wealth of molecular biological and biochemical evidence supporting these models, the mitochondria of today are best seen as descendants of cells that evolved within other cells (Gray, 1983).

A similar story may be told of plastids, the photosynthetic inclusions in all plant and algal cells. Plastids probably originated as free-living, photosynthetic prokaryotes. Green *Prochloron* and cyanobacteria were eaten by protists with mitochondria already in place. But not all were digested and some survived symbiotically within their hosts. Such liaisons led on the one hand to the green algae and plants (*Prochloron* bacteria became chloroplasts) and on the other to red seaweeds (cyanobacteria such as *Synecococcus* became rhodoplasts). The molecular biochemical evidence for a common lineage between photosynthetic bacteria and photosynthetic organelles, especially rhodoplasts, is even stronger than that for a common lineage between mitochondria and respiring bacteria (Gray and Doolittle, 1982).

**Fig. 19.** Microbes of termite hindguts. A. Electron micrograph of bacteria (b) surrounding two protists (p) in the hindgut of a wood-ingesting termite (bar = 1 micrometer). Note the layer of bacteria surrounding the surface of the protist at the left. B. Scanning electron micrograph accurately portrays the density and diversity of the termite hindgut microcosm (bar = 10 micrometers). (Courtesy of Dr. David Chase.)

## NEW CELLS FROM MICROBIAL COMMUNITIES

Clones of eukaryotic cells in the form of animals, plants, fungi, and protoctists seem to share a symbiotic history. Protoctists, the most varying and the earliest to evolve of all these groups, are defined as all protists and their multicellular descendants, exclusive of animals, plants, and fungi (fig. 20). Examples include the brown seaweeds and the *Acrasia* slime mold. No protoctists form embryos, nor are any able to complete their life cycles on land. Thus protoctists are considered the most ancient of eukaryotes, preceding the spore-forming fungi and the embryo-forming plants and animals. From an evolutionary point of view, the first eukaryotes were loose confederacies of bacteria that, with continuing integration, became recognizable as protists, unicellular eukaryotic cells. Loose intracellular symbioses between different forms of bacteria, such as parasitic oxygen-respirers (protomitochondria) and larger acid- and heat-resistant host forms (protonucleocytoplasm; Searcy, Stein, and Green, 1978; Margulis, 1981) eventually tightened and formed cells. The earliest protists were likely to have been most like bacterial communities (Margulis, Chase, and Guerrero, 1985). At first each autopoietic community member replicated its DNA, divided, and remained in contact with other members in a fairly informal manner. *Informal* here refers to the numbers of partners in these confederacies: they varied. As long as at least one member of each type (respiring protomitochondrion and its host) stayed in contact, natural selection acted to preserve the respiring, resistant complex. The emergence of eukaryotic cells from bacterial communities is diagramed in figure 21, which shows the relationship of the major organelles to present-day bacteria, with which they are thought to have common ancestors.

Traditionally, protists have been labeled "miniature plants" if they contain plastids and photosynthesize, "animalcules" if they lack plastids, require ready-made food, and are mobile. The existence, however, of protists such as *Euglena,* which, although photosynthetic, also dart around in the mobile fashion of animals, has long shown the inherent fallacy of dividing all organisms into *either* plant or animal. Early eukaryotic microorganisms and their modern descendants, such as amoebae, ciliates, green algae, diatoms, dinomastigotes (dinoflagellates), and red, brown, and yellow-green seaweeds, are now best considered in their own large group: kingdom Protoctista (Margulis and Schwartz, 1982). There are estimated to be over 100,000 species of protists and their multicellular descendants in the kingdom (Corliss, 1984). The recognition that protoctists are legitimately neither animal nor plant but distinctive organisms derived ultimately from coevolved communities of bac-

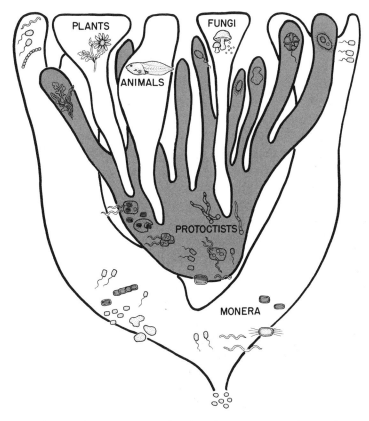

**Fig. 20.** Relation of protoctists to other kingdoms. (Drawing by Laszlo Meszoly.)

teria is crucial for the argument of this book. It is in protoctists that the meiosis-fertilization type of sexuality evolved. Although there is disagreement about details, biologists agree that mitosis evolved in protists following a complex, tortuous path (Roos, 1984). It is not possible to understand the origin of the various steps in the emergence of meiotic sexuality without a prior knowledge of the nature of the protoctist cell and its mitotic systems.

## PROTOCTISTS

Because certain protoctists have been so important in human history a bit of information is available about some of them, although vast numbers of species remain unknown to science. Some, like the malarial parasite, a member of the genus *Plasmodium*, have entire libraries dedicated to them, while

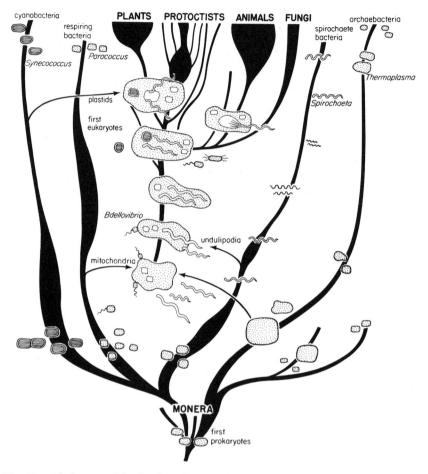

**Fig. 21.** Phylogeny of the five kingdoms. Scheme showing the relation of organelles to bacteria, from which they are thought to have evolved. (Drawing by Laszlo Meszoly.)

others, like the xenophyophores, deep-sea organisms that produce barium sulfate skeletons in some unknown way, have never even been seen alive. Our purpose here is only to provide a brief introduction to this astounding group of microbes and their descendants that seem to have invented everything distinctive about the eukaryotic world (for example, meiosis, differentiation, and programmed death). With the exception of developing semes (genetic complexes) related to habitats on dry land, such as bones, lignified cell walls, wings, and language, protoctists are able to do whatever animals and plants can do—and more.

Knowledge of protoctists has been limited by the traditional ways in which they have been studied. Scattered in the literatures of zoology, botany, and mycology, general information about protists is difficult to obtain. Even today there are arguments about whether the nature of these organisms is primarily plant, fungal, or animal. It is none of these. Those of us who study this extraordinarily variable kingdom are beginning to recognize over thirty distinctive groups (phyla) of protoctists (fig. 22). We expect fundamentally different new phyla of smaller protoctists to be revealed by natural history and electron microscopy within the next decade or so.

Although protoctists are defined as eukaryotes excluded from the animal, plant, and fungal kingdoms, they have positive identifying features as well. Protoctists are aquatic organisms that vary from the size of bacteria (for example, *Nanochlorum*, the smallest photosynthetic protist known, is a little larger than one micrometer in diameter; Zahn, 1984) to giant seaweeds a hundred meters long. Unlike animals, fungi, and plants, protoctists may lack mitosis, microtubules, mitochondria, and other common eukaryotic organelles. The distribution, including the absence, of these features considered essential to the sexual meiotic process provides us with fascinating clues to its origins. All protoctists lack embryos and embryonic development, major features in animals.

Except for a few groups, members of a given protoctist phylum tend to have characteristic cell structures, including kinetids. Kinetids (fig. 23) are unit structures that always include at least one kinetosome (fig. 24). They consist of microtubules and fibers arranged in patterns surrounding the one or more kinetosomes at the bases of undulipodia.

Nutritionally protoctists are heterotrophs, feeding by absorption (osmotrophically), by ingestion (phagotrophically), or they are photosynthetic. They are never chemoautotrophs. All derive their energy from organic compounds or sunlight. Strict photoautotrophy is rare; nearly all require some organic constituent from their medium. For example, dinomastigotes and euglenids (euglenas) grow photosynthetically but need a supplement of tiny quantities of vitamin $B_{12}$. Some protoctist cells are without walls; some have cellulosic, proteinaceous, chitinous, or other types of walls. They may develop elaborate cell structures composed of microtubules and yet be entirely without sexuality. For example, the two great groups of radiolarians (spumellarians and phaeodarians)—probably not directly related to each other—both make elaborate skeletal spines and actinopods (axopods) underlain by microtubules. No sex, however, is known in any member of these two groups. On the other hand, some foraminifera may have incredibly complicated sexuality accompanied by distinctive sexual stages.

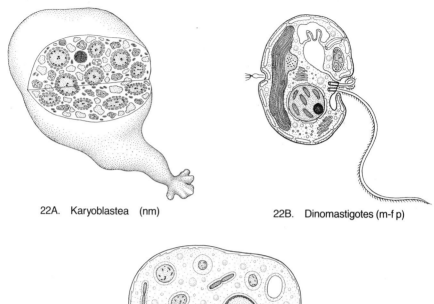

22A.  Karyoblastea  (nm)          22B.  Dinomastigotes (m-f p)

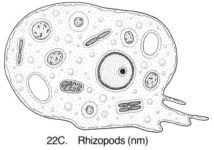

22C.  Rhizopods (nm)

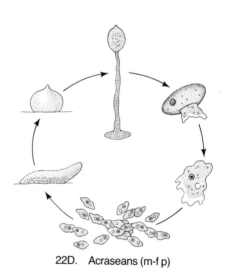

22D.  Acraseans (m-f p)

**Fig. 22A–II.**   Major phyla of the kingdom Protoctista. m-f = meiosis-fertilization documented; m-f p = meiosis-fertilization possible; nm = no meiosis-fertilization. (Drawings of AA and FF by Christie Lyons, left portions of HH and II by Emily Hoffman, all others by Laszlo Meszoly.)

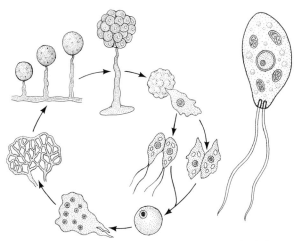

22E.   Amoebomastigotes and Myxomycotes (m-f p)

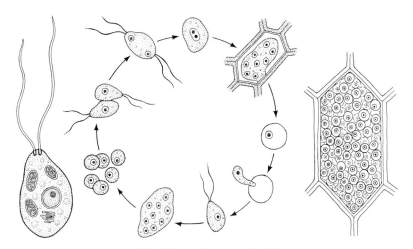

22F.   Plasmodiophorans (m-f p)

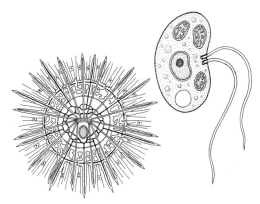

22G.   Actinopods (m-f p)

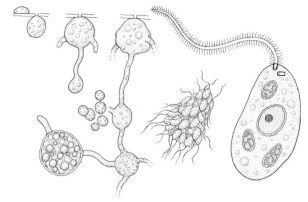

22H.   Hyphochytrids (nm)

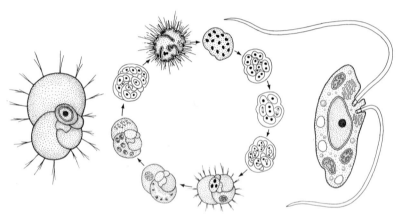

22I.   Foraminifera (m-f)

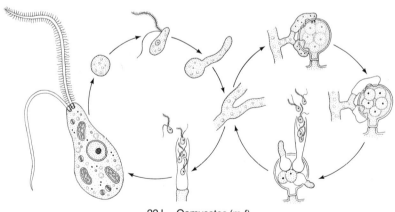

22J.   Oomycotes (m-f)

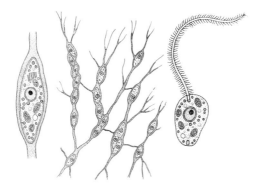

22K.    Labyrinthulids and Thraustochytrids (m-f p)

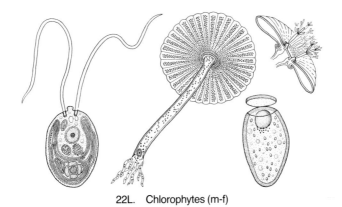

22L.    Chlorophytes (m-f)

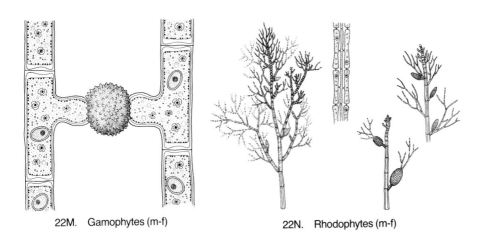

22M.    Gamophytes (m-f)

22N.    Rhodophytes (m-f)

79

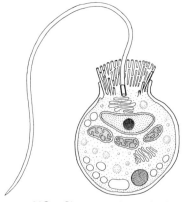

22O.  Choanomastigotes (nm)

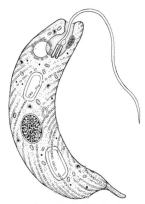

22P.  Euglenids (nm)

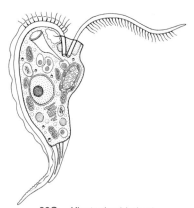

22Q.  Kinetoplastids (nm)

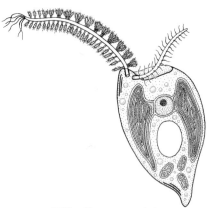

22R.  Chrysomonads (nm)

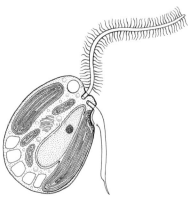

22S.  Xanthophytes (m-f p)

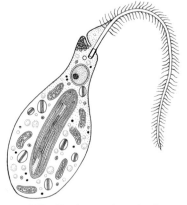

22T.  Eustigmatophytes (nm)

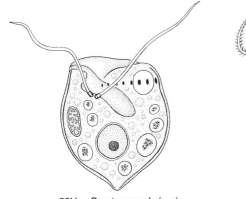

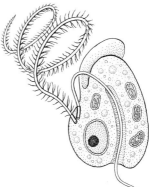

22U.   Cryptomonads (nm)

22V.   Bicoecids (nm)

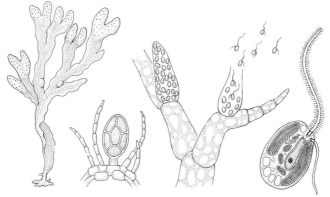

22W.   Phaeophytes (m-f)

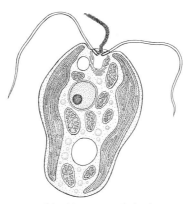

22X.   Haptomonads (nm)

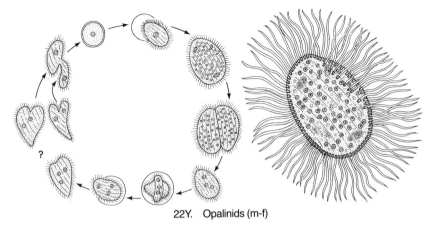

22Y. Opalinids (m-f)

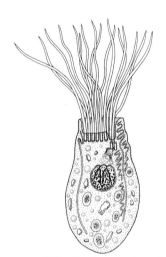

22Z. Parabasalids (m-f p)

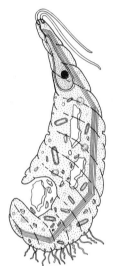

22AA. Metamonads (m-f p)

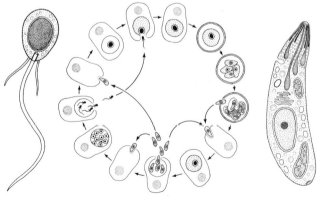

22BB. Apicomplexa (m-f)

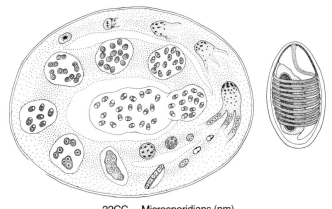

22CC. Microsporidians (nm)

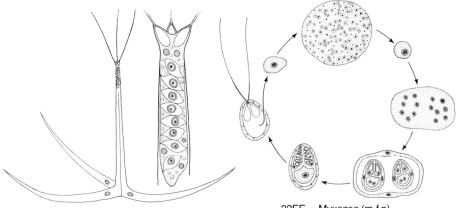

22DD. Actinomyxids (m-f p)  22EE. Myxozoa (m-f p)

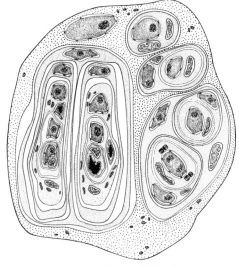

22FF. Paramyxids (nm)

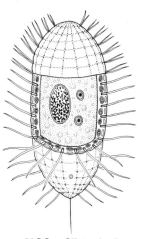

22GG. Ciliates (m-f)

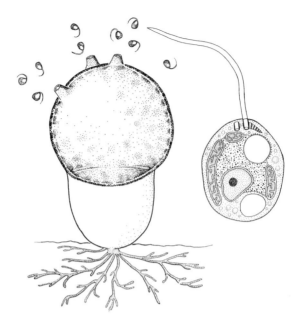

22HH.   Chytrids (m-f)

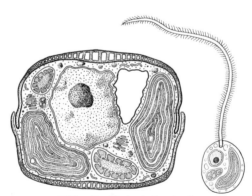

22II.   Bacillariophytes (m-f)

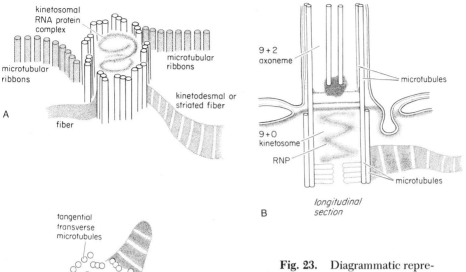

A

kinetosomal
RNA protein
complex

microtubular
ribbons

microtubular
ribbons

kinetodesmal or
striated fiber

fiber

9 + 2
axoneme

microtubules

9 + 0
kinetosome

RNP

microtubules

B    *longitudinal section*

tangential
transverse
microtubules

9 + 0
kinetosome

microtubules

C    *transverse section*

**Fig. 23.** Diagrammatic representation of a kinetid in three-dimensional, longitudinal, and transverse section. Although generalized, these types of kinetids are characteristic of ciliates. (Drawing by Laszlo Meszoly.)

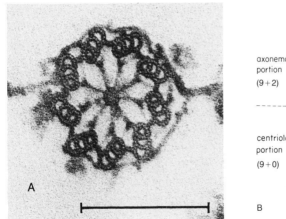

A

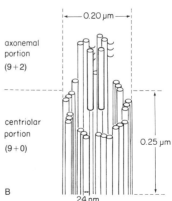

axonemal
portion
(9 + 2)

centriolar
portion
(9 + 0)

0.20 μm

0.25 μm

24 nm

B

**Fig. 24.** Kinetosomes. A. Electron micrograph of a kinetosome from *Trichonympha* (bar = 0.25 micrometers). (Courtesy of R. W. Linck, Harvard Medical School.) B. Drawing showing dimensions of kinetosomes, which are constant regardless of species. (Drawing by Laszlo Meszoly.)

85

## THE EARLY PROTOCTISTS

The early evolution of protoctists cannot be traced directly in the fossil record. One possible representative of the most ancient of these creatures, the nonmitotic giant amoeba *Pelomyxa palustris*, lives in freshwater lakes and ponds (Whatley and Chapman-Andresen, n.d.). Because it does not form hard parts it simply disintegrates when it dies, becoming food for bacteria and plant roots. The fossil record is very weak in general on the evolution of watery protoctists. Inferences concerning their evolution are usually made by physiological and electron microscopic comparisons among modern forms.

Some protoctists show significant variations on the theme of mitosis. For example, the dinomastigotes (dinoflagellates) lack typical chromosomes, but in dividing they retain connections of "naked" DNA (not coated with histone protein) with the nuclear membrane. Protoctist cell divisions vary from direct division of nuclei not resembling standard mitosis at all, in *Pelomyxa palustris* (Daniels and Breyer, 1967), through the peculiar mitoses of dinomastigotes (Herzog, Van Boletsky, and Soyer, 1984), to typical plant-animal mitosis. The logical explanation of these observations is that mitosis evolved in protoctists and was fine-tuned and stabilized in some lineages. Selected lineages, already mitotic, led further to fungi, animals, and plants, as well as to other meiotic protoctists. Many protoctist lineages failed to evolve the meiotic fertilization cycle. The decendants of such nonmeiotic lineages are still quite common (for example, amoebae, euglenids, trypanosomes) and their study illuminates the origin of mitosis itself. A stable mitotic division cycle in a lineage is an absolute prerequisite if meiosis is to evolve.

Since eukaryotes are aerobic organisms, already adapted to atmospheric oxygen, and are morphologically more elaborate than prokaryotes, no serious scholars have argued that bacteria, which clearly had anaerobic ancestry, arose from eukaryotes. Many evolutionary biologists have suggested, however, that nuclei and other eukaryotic organelles (mitochondria and plastids) are the result of a gradual evolution from bacteria of a type of cell essentially distinct from the bacteria. This view has many names: the autogenous theory, direct filiation of the origin of eukaryotes, the differentiation theory, the classical theory, or the nonsymbiotic theory.

In the nonsymbiotic theory of the origin of eukaryotic cells, bacterial cells first developed internal membranes, some of which packaged the cell's RNA and DNA in a separate nucleus. Then the organelles in question "pinched off"; the nuclear genes escaped from the nucleus and enveloped themselves in membranous material, forming the DNA-containing chloroplasts of green photosynthetic protists and the DNA-containing mitochondria we see today in all protoctist lineages. Eventually, according to this theory, another version of which is called the "cluster-clone hypothesis" (Bogorad, 1975), the genes that escaped from the nucleus became the genes of plastids and mitochondria. In any version of the autogenous theory of organelles, gene products in the nuclei and in organelles should, due to their common descent, be more similar to each other than they are to free-living bacteria. Similarities between nucleocytoplasm and organelles are really rather rare, however, and organellar gene products tend to be more like those of appropriately chosen free-living bacteria than like the nucleocytoplasm in which they reside.

Thus the autogenous theory has several shortcomings when compared to "SET," the abbreviated name given to the serial endosymbiosis theory of protoctist origins. A major deficiency is that the autogenous theory does not explain the fact that the DNA and proteins of mitochondria are extremely different from those of the surrounding cytoplasm. The molecular biology of plastids inside algal cells is far more similar to that of cyanobacteria than it is to that of the nucleocytoplasm surrounding the plastids. The genetic material of these organelles consists of free-floating genophoric strands not bound with histone proteins into nucleosomes. Furthermore, this organellar DNA replicates out of synchrony with respect to the cell's nuclear DNA. The several kinds of organellar RNA found in plastids (messenger, ribosomal, and transfer RNA) are easily understood as residual from an original free-living status. The RNA and proteins of mitochondria are quite similar to the RNA and proteins of certain respiring bacteria such as *Paracoccus denitrificans*, because *Paracoccus* and mitochondria have more recent common ancestors than do mitochondria and nucleocytoplasm.

The behavior of plastids and mitochondria, membrane-bounded DNA- and RNA-containing organelles that grow and divide in the cytoplasm of the cell independently of the nucleus, resembles the nonmitotic mode of free-living bacterial reproduction. No organelle except the nucleus ever forms chromosomes and divides by mitosis. SET, which postulates a symbiotic origin for these cytoplasmic organelles, explains many aspects of the extraordinary differences between eukaryotic and prokaryotic cells: they have different origins. Prokaryotes are single genomes, true units. Eukaryotic organelles, on the other hand, began as prokaryotes. Thus eukaryotes are complexes of several heterologous units, each class of organelle originally stemming from a different bacterial type. The fundamental concept that eukaryotic cells originated from the merging of confederacies of interacting bacteria of very different origins is essential to our analysis of the origin of mitosis and the subsequent development of meiotic sexuality. A further critical assumption is that symbiosis, the living together of organisms of different species, was a crucial prerequisite in the evolution of still another class of organelles, the microtubular organizing centers (MTOCs). This too is indispensable to our understanding of the origin of meiotic sex.


## PROTOCTIST DIVERSITY


The protoctists comprise all organisms excluded from the other three kingdoms of eukaryotes. They are neither animals (they do not develop from blastulas), nor plants (they do not develop from embryos supported by sterile tissue), nor fungi (haploid or dikaryotic mycelial organisms that develop from spores); and of course they are not bacteria. The protoctists are an immense group that includes eukaryotic microorganisms and their immediate multicellular descendants. Electron microscopic, genetic, and biochemical studies since the 1960s have revealed a diversity of structure and life-styles in the microcosm of protoctists that was unanticipated from the perspective of the macroscopic world.

The best-known protoctists are called "algae" (or "eukaryotic algae," since the blue-greens are prokaryotes) and "protozoa." The presence of plastids in very closely related organisms converts a "protozoan" to an "alga." The more that is learned about the protoctists, the more these terms become obsolete. They should be used only informally, to refer to nutritional modes (algal: photosynthetic; protozoological: heterotrophic) rather than to evolution and systematics. In addition to "protozoa" and "algae" the protoctists include groups that have traditionally been placed with the fungi (water molds, chytrids, slime molds) and with the plants (seaweeds). Distinctions among the protoctist groups are made on the basis of morphology and life-cycle stages, presence or absence of organellar systems (mitochondria, undulipodia), morphology of organellar systems, especially the kinetosomal-microtubule arrangement at the base of the undulipodia, which, by definition, are kinetids (Moestrup, 1982). Mitosis and meiotic sexuality may be absent. In the sexual species the nature of the sexual stages is crucial for the differentiation of the groups. The following brief description of protists is meant to accompany the diagrams on the following pages and is by way of an introduction to the protoctist kingdom of living organisms. Only recently has comprehensive information about protoctists been available in one source, *The Protoctista* (Margulis, Corliss, and Chapman, n.d.), comparable to those for bacteria: Bergey's manual (Buchanan and Gibbons, 1974) and more recently *The Prokaryotes* (Starr et al., 1983). We summarize here the most recently recognized list of monophyletic higher taxa in full recognition that with further research the groupings will change. Our short description must suffice: details may be found in *The Protoctista*.

Because of the confusion of the terms *flagella* and *flagellate*, which designate protoctist phyla but also refer to completely different bacterial structures made of flagellin proteins, we use here the term *mastigote*. Coined by protozoologists, this word has been limited to eukaryotes and thus is used here to mean a motile, "flagellated" eukaryotic cell. We use *mastigote* to refer to cells that bear undulipodia for motility. If mastigote cells need fusion with eggs to continue to grow, they are, by definition, sperm (fig. 22A); if they develop on their own, by definition they are zoospores.

### Karyoblastea (amitotic amoebae)

The phylum Karyoblastea includes anaerobic giant amoebae that lack mitochondria, chromosomes, mitotic spindles, mitosis, and meiotic sex. Several types of bacterial endosymbionts have been noted inside karyoblastea, including methanogens, which are anaerobic bacteria capable of producing methane gas. *Pelomyxa palustris*, a pond-water form, is virtually the only species studied. Undulipodia may be present on its surface.

### Dinomastigotes (dinoflagellates)

Several thousand species of dinomastigotes, freely floating (planktonic) microorganisms, are known, most of them marine. They have a characteristic distinctive morphology with one transverse "girdle" and one longitudinal undulipodium. Some dinomastigotes whirl (*dinos* means "rotate" in Greek), accounting for their name (fig. 22B). Mitochondria and generally two undulipodia are present. The DNA organization is unique in this group. Histones forming nucleosomes are absent. The unit DNA fibril densely packed to make permanently condensed chromosomes is more like that of prokaryotes than like the nucleosome-studded particle of other eukaryotes (Herzog, Von Boletsky, and Soyer, 1984). Hydroxymethyl uracil is a very common replacement for thymine and it is incorporated into dinomastigote DNA during synthesis (Galleron, 1984). Both photosynthetic and heterotrophic modes of nutrition are present. Pairing and fusion to form resistant cysts is common—it can even be induced in the laboratory—but standard meiotic processes have not been documented.

### Rhizopods (amastigote amoebae)

Rhizopods—ubiquitous heterotrophic marine and freshwater amoebae—lack kinetids and undulipodia at every stage in their life cycle. Some lack mitochondria and microtubules. Sex is absent. Movement is by pseudopods. Many rhizopods form desiccation-resistant cysts (fig. 22C).

### Acraseans (amastigote cellular slime molds)

Acraseans or cellular slime molds are heterotrophic soil microbes that lack kinetids and undulipodia at all stages. Electron micrographs of thin sections of their mitochondria show the internal membranes to be tubular, rather than flat like those of animals. Meiotic sexuality has not adequately been demonstrated. These amoebae come together to form motile, multicellular, aggregate "slugs" that differentiate: the cells change in form and function without undergoing cell division. The aggregate converts into stalked dissemination structures, sorocarps, composed of hundreds of cells (these have confusingly been named "fruiting bodies" or "sporangia"), which release spherical, resistant, thick-walled cysts (that have confusingly been called "spores"). When the cysts germinate, amoebae, capable of aggregation again into slugs, emerge from them (fig. 22D).

### Amoebomastigotes and Myxomycotes

Amoebomastigotes (amoeboflagellates) and myxomycotes (also known as eumycetozoa, mycetozoa, acellular slime molds, plasmodial slime molds, true slime molds, myxomycotina) are heterotrophic microbes with feeding mononucleate and multinucleate (plasmodial) stages. A plasmodium (plural: *plasmodia*) can be thought of as a large, multinucleate amoeba. Species in this group have anteriorly directed undulipodia with characteristic kinetids: one striated (striped) rootlet and one kinetosome having an anteriorly directed undulipodium and associated with four or five sets of microtubules. The cells of members of this group transform from amastigote amoebae, with no undulipodia or kinetids at all, to mastigote stages by the appearance of kinetosomes and their kinetids and growth of undulipodia (fig. 22E). They transform back to the amoeba stage by active resorption (withdrawal) of their undulipodia. They are capable of forming desiccation-resistant cysts. Meiotic sex is absent in unicellular forms. Cell fusion has been reported—and claimed to be meiotic sex in some of the multicellular forms—in the plasmodial slime mold species; at least in *Echinostelium* this is doubtful (Haskins and Therrien, 1978).

### Plasmodiophorans

Plasmodiophorans, which are endoparasitic multinucleate (plasmodial) heterotrophs, are primarily found in soil and plant tissue. Cysts germinate to form undulipodiated cells, or mastigotes. Although they look like sperm they are not, because they develop without having to fuse with an egg first. The mastigotes develop to form plasmodia. From plasmodia, undulipodiated gametes form and fuse. Diploid zygotes develop into a plasmodial stage again, usually inside a host plant (fig. 22F).

### Actinopods (radiolarians, acantharians, heliozoans)

Actinopods, large for the microscopic world, are skeletalized hetero-trophs characterized by long cell processes (protrusions) called "axopods." Axopods, containing structures based on microtubules that often display highly complex geometrical patterns, are used for feeding and sometimes for locomotion or buoyancy. Actinopods include marine forms—radiolarians (spumellarians and phaeodarians) with calcium carbonate skeletons and acantharians with strontium sulfate skeletons—and freshwater forms—he-liozoans with silica skeletons. Fusion of products of mitosis within a common cyst interpreted to be sex has been observed in heliozoans. Sexuality is un-known in the other groups. Propagation by spermlike mastigote forms has been reported. No marine forms and very few freshwater forms have been cultivated in the laboratory (fig. 22G).

### Hyphochytrids

Hyphochytrids resemble the white threads of fungi in wet soil or pond water. They are freshwater heterotrophic multicellular organisms that pro-duce mastigotes with characteristic kinetid patterns. Mastigote stages repre-sent the only way in which these organisms can reproduce. Sexuality is ab-sent. Mastigotes round up and develop into long threads, hyphae, that form multicellular walled structures from which a plasmodial form emerges. Mas-tigotes then develop from this plasmodial structure by differentiation and swim away (fig. 22H).

### Foraminifera

Foraminifera are marine heterotrophic organisms. They have complex calcium carbonate skeletons and elaborate sexual stages involving mastigote or amoeboid male and female gametes that fuse. Some forams have two sorts of nuclei: one generative, capable of further growth by division, and one somatic, which functions in cell processes but fails to divide. The largest forams have multichambered skeletons and attain a size easily seen with the naked eye, even though they are single (multinucleated) cells (fig. 22I).

### Oomycotes ("water molds")

The oomycotes are heterotrophic, funguslike protoctists. They undergo sexual conjugation. Hyphae of the opposite mating types fuse and grow new hyphae. Eventually mastigote stages develop and are released from the hyphae. The mastigotes, by which oomycotes are disseminated, have characteristic kinetids with one forward-directed undulipodium and a second distinguishing trailing one. This cell shape is called "heterokont" (fig. 25). It is frequently encountered in protoctist groups and may indicate an underlying common evolutionary history in oomycotes, brown algae, chrysophytes, and other groups that display it (fig. 22J).

### Labyrinthulids and Thraustochytrids (slime nets, labyrinthulomycotes)

More commonly known as slime nets, the labyrinthulids are colonies in which cells travel inside an extracellular matrix composed of polysaccharide and actin proteins (like those in muscles). Labyrinthulids have heterokont mastigote stages, and may distribute themselves by zoospores. Thraustochytrids thought to be chytrid fungi are now recognized as relatives of the labyrinthulids (fig. 22K). They, too, form extracellular matrices in which cells are embedded; in addition they form an enclosing chytrid-like structure from which heterokont zoospores emerge. These marine organisms are related because they produce the extracellular matrix and its outer membrane using specialized structures called sagenogenosomes (identical with bothrosomes) embedded in the cell membranes (Porter, 1986).

### Chlorophytes (green algae)

Chlorophytes are the most common green algae. Hundreds of species of multicellular and unicellular haploid photoautotrophs (photosynthesizers) are included in this phylum. Chlorophytes form sexual stages by making undulipodiated gametes that fuse. The fused diploid zygote forms a structure that resists desiccation and in which meiosis occurs. The volvocales, ulvales, charales, caulerpales, siphonales, and others belong to the chlorophytes, from microscopic to large marine and freshwater algae. Land plants most likely evolved from such algae (fig. 22L).

### Gamophytes (conjugating green algae)

The gamophytes are green unicellular or filamentous algae that never form kinetids or undulipodia. In their sexual processes hyphae of opposite

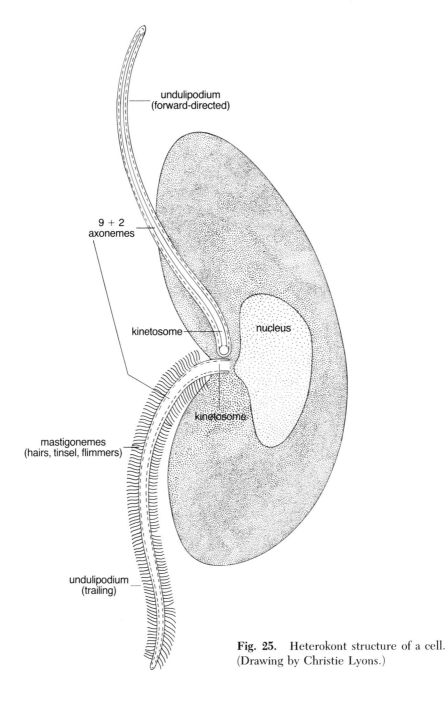

undulipodium
(forward-directed)

9 + 2
axonemes

kinetosome

nucleus

kinetosome

mastigonemes
(hairs, tinsel, flimmers)

undulipodium
(trailing)

**Fig. 25.**   Heterokont structure of a cell.
(Drawing by Christie Lyons.)

mating types conjugate to form a diploid zygote. The zygote undergoes meiosis to regenerate haploid cells, which grow into filaments or lovely delicate water organisms called "desmids" (fig. 22M).

## Rhodophytes (red algae)

Rhodophytes (Greek, "red plants") include all the red algae, the vast majority of which live in the sea. There are hundreds of species. Those studied have well-developed mitosis and sexuality: male spermlike cells alight on females and fuse with them. However, no kinetids or undulipodia (mastigote stages) are formed; the sperm are not motile but carried passively in the water (fig. 22N).

## Choanomastigotes (choanoflagellates) and Craspedomonads

Choanomastigotes are heterotrophic bacteriovores, craspedomonads are photosynthetic; both are collared single cells and colonial forms. Their cells, which have the same kind of complex kinetids, are very similar to each other, as well as to the cells of porifera (sponges), which are classed as animals. For this reason sponges are thought to belong to a different original lineage from that of the rest of the animals. Sex has not been adequately documented in the group. Cells have characteristic kinetid structure (fig. 22O).

## Euglenids (euglenophytes) and Kinetoplastids

Euglenids (fig. 22P) are undulipodiated, motile single cells and colonies. They exist primarily in a free-living and photosynthetic state, although some euglenids lack plastids. Kinetoplastids are small heterotrophic mastigotes, including parasitic forms such as the trypanosomes. Kinetoplasts, which give the group its name, are membrane-bounded organelles found in many kinetoplastids (fig. 22Q); they contain DNA and lie near the kinetosome and its undulipodium. At least in part the kinetoplast DNA is modified mitochondrial DNA. The kinetoplast of kinetoplastids can be envisaged as a highly specialized mitochondrion. Euglenids and kinetoplastids have characteristic kinetids. Their mitosis is peculiar and sexuality is absent in these groups.

## Chrysomonads (chrysophytes, golden yellow algae)

The chrysomonads include golden yellow algae and some related heterotrophs. The mastigote stages are heterokont in form. Hundreds of different colonial species have been described. The colonies usually reproduce irregularly. A piece of the colony breaks off, is washed away, and starts growing on its own by mitotic division of its component cells. No kind of sexuality has ever been described in chrysomonads (fig. 22R).

### Xanthophytes (yellow-green algae)

Xanthophytes are freshwater algae. The cells are in the form of heterokont mastigotes, and their multicellular derivatives form filaments or sheets. Each cell contains yellow-green plastids with a distinctive set of chlorophyll pigments. No sexuality has been described for any members. They form cysts impregnated with metal or other substances, which survive cold and desiccation (fig. 22S).

### Eustigmatophytes and Raphidiophytes ("chloromonads")

Like xanthophytes, eustigmatophytes and raphidiophytes are small freshwater mastigote algae with characteristic cell structure including a single kinetosome and undulipodium underlain by a distinctive kinetid. The cells harbor yellow-green plastids and no sexuality of any kind has ever been documented in them. Only a small number of species has been described, since electron microscopic studies are required to distinguish these little mastigote algae from xanthophytes, chrysomonads, and the like (fig. 22T).

### Cryptomonads (cryptophytes)

Cryptomonads are very common microorganisms, single cells found in marine, brackish, and fresh water. These little mastigotes have deep grooves in their cells, associated with which are kinetids with kinetosomes from which emerge undulipodia, generally two. Some cryptomonads have plastids and are photosynthetic, others eat bacteria or dissolved nutrients. Rather than by standard mitosis, these mastigotes divide by an extensive developmental process that involves their undulipodia and the formation of a groove. Sex has never been seen in any of them (fig. 22U).

### Bicoecids

Bicoecids are small, heterokont, heterotrophic, bacteriovorous mastigotes primarily found in soil. They have characteristic kinetids. In their small number of species no sexuality has ever been reported (fig. 22V).

### Phaeophytes (brown algae)

A major group of protoctists, the phaeophytes are brown algae, including kelps and all brown seaweeds (fig. 22W). Mitosis and meiotic sexuality is well established here, and many large haploid and diploid, plantlike structures are known. Mastigote stages are found associated with sperm only. The sperm are heterokont cells with two undulipodia like the cell drawn in fig. 25.

### Haptomonads (coccolithophorids, haptophytes, prymnesiophytes)

Coccolithophorids are single-celled and colonial, marine mastigotes covered with scales: calcium carbonate skeletal plates (called "coccoliths"). Coccolithophorid, haptomonad, or haptophyte cells (these are all synonyms) bear a characteristic microtubule-based structure, a haptoneme, which serves as a holdfast. It has 6 + 0 instead of 9 + 2 microtubules, but this structure too develops from an MTOC. Coccolithophorids are primarily photosynthetic, marine, planktonic microorganisms. Sexuality has never been described in any of the many species that are members of the group (fig. 22X).

### Opalinids

Opalinids are heterotrophic microbial parasites that live in the digestive tracts of amphibians. Electron microscopy has determined that they have characteristic kinetids arranged in rows on their surfaces. Sexuality involves fusion of undulipodiated gametes and the formation of cysts. Many species have been seen but hardly any are known in detail (fig. 22Y).

### Parabasalids (trichomonads, hypermastigotes)

The parabasalids include very complex mastigotes symbiotic in the guts of insects: devescovinids, trichomonads, and hypermastigotes. They are heterotrophs lacking mitochondria and bear from about four to more than a hundred thousand undulipodia, depending on species. The undulipodia are underlain by complex kinetids. Sexuality has been described in a few species. The study of sexuality in these mastigotes, such as the hypermastigote *Barbulanympha* (primarily by L. R. Cleveland), has been crucial for understanding meiotic sexuality in general. Unfortunately, the total number of studies of sexuality in this group is very small. The parabasal body, for which the group is named, is a modified Golgi apparatus (fig. 22Z).

## Metamonads (oxymonads)

Metamonads are microscopic, often parasitic, heterotrophic mastigotes. Some are found in soil and fresh water. Included are the diplomonads, oxymonads (pyrsonymphids), and polymonads. They have characteristic complex kinetids associated with their nuclei and complex microtubules and microfibrils. They lack mitochondria. Fusion of gametes and sexual cycles have been reported in some species (fig. 22AA).

## Apicomplexa ("sporozoans")

The apicomplexa are all parasites; they are heterotrophs with characteristic cell structures at the end of the cell called the "apical complex." This set of unique cell modifications permits the penetration of animal tissue. At the anterior end of the cell they have conspicuous dense structures called "rhopteries," which are associated with mitochondria. This part of the apical complex can be seen with the light microscope. Mastigote stages are limited to the male gamete, which fuses with the larger female gamete. Reproduction is by multiple fission (nuclear mitoses in which several nuclear divisions precede cytoplasmic divisions), and small cells are formed (fig. 22BB).

## Microsporidians (microsporidian parasites)

Microsporidians are heterotrophic, parasitic microorganisms the cells of which have a characteristic structure, the polar filament, for injection of cytoplasm (sporoplasm) into their animal hosts. Parasitic microsporidians also grow by multiple fissions in which several nuclear divisions precede cytoplasmic divisions. They lack mitochondria and even kinetids and undulipodia are unknown. Sexuality has not been documented in the group, and their ribosomes, resembling those of prokaryotes, are smaller than those in the cytoplasm of other eukaryotes (fig. 22CC).

## Actinomyxids (actinomyxid "sporozoans")

Actinomyxids are heterotrophic parasites that attack annelids. Kinetids are unknown in them. Sexuality is incompletely documented (fig. 22DD).

## Myxozoa (myxosporidian "sporozoans")

Myxozoa are multicellular parasites in which cells form inside cells. The organisms produce multinucleate resistant reproductive structures confusingly called "spores." From the "spores" grow out filamentous parts of cell modified for anchoring to host animal tissue. Kinetids, kinetosomes, and undulipodia, as well as any form of sexuality, are incompletely documented (fig. 22EE).

### Paramyxids

Members of Paramyxea such as *Marteilia* and *Paramyxa* are all parasitic in marine animals. Paramyxids appear as infections in polychaete worms, mollusks, and crustaceans. A stem cell, such as shown in fig. 22FF, produces inside itself "sporonts" which produce more cells inside themselves by an internal kind of cell division. This arrangement of a cell enclosed inside another cell is unique. Centrioles in these organisms have nine singlet, instead of the usual triplet, tubules.

### Ciliates (ciliophora)

The ciliates, with their characteristic kinetids, are sexual heterotrophic unicells with dimorphic nuclei: diploid meiotic micronuclei and nonmitotic somatic macronuclei. Named for their abundance of surface undulipodia, nearly 10,000 species are estimated to exist (fig. 22GG).

### Chytrids (thallus)

The chytrids are heterotrophic, multicellular diploids easily seen with the naked eye. Their activities resemble fungal growth and they are found in watery places such as decaying vegetation. They propagate only by the formation of mastigote stages that look like little sperm and have been called "zoospores." The mastigotes have characteristic kinetids. The kinetosomes form posteriorly directed undulipodia. The kinetids with their undulipodia can be resorbed into the cell. That is, the zoospore retracts its undulipodia and begins to form the sacklike chytrid body when food is sufficiently abundant. The chytrid body, with its chitinous walls and thready, rootlike structures, is the part of the organism that is generally studied. Many species are known, primarily in fresh water or soil or parasitic in plants. Sexuality is common in the group (fig. 22HH).

### Bacillariophytes (diatoms)

There are thousands of diatom species. All are photosynthetic, and most form delicate, symmetrical, lacy skeletons of silica. Most species are marine.

Sexuality is common; the ordinary diatom is the diploid product of fertilization of two unequal gametes (anisogamy). The undulipodiated stage exists only in the sperm. The sperm-tail undulipodium, like that of animals and chytrids, is posteriorly directed. Meiosis occurs prior to the formation of gametes (fig. 22II).

## MICROTUBULE ORGANIZING CENTERS

The MTOCs, sometimes abbreviated as MCs (*microtubular centers*), are more a brilliant concept devised by Jeremy Pickett-Heaps (1971) than a distinct and constant morphological entity. Many years of observation, beginning in the nineteenth century and summarized by E. B. Wilson (1925), established the idea of centers of organization in eukaryotic cells, especially in large protists (such as heliozoans) and egg cells. These centers seemed to be the source of the development of certain characteristic cell structures: axopods, cilia, asters, mitotic spindles, and so forth grew from these centers at specific times in the cell cycle.

The innovations of electron microscopy, beginning with the introduction of glutaraldehyde as a fixative around 1962, greatly clarified these observations. The centers of activity, appearing granulated or fibrous and often fuzzy, were places of organized deployment of long, thin, hollow cell structures called "microtubules." The tubules, amazingly constant at 240 angstroms in diameter and widely varying in length, were seen to be the structural basis of many familiar cell organelles. Their constant diameter and peculiar patterns undoubtedly indicate a common ancestry for certain microtubule-based organelles such as cilia and sperm tails. There is still no consensus among biologists concerning the origin of microtubule-based organelles except for the idea that they share a common ancestry, whatever it may be.

Our theory that microtubule-based organelles such as cilia evolved symbiotically, like mitochondria and plastids, will immediately suggest how these structures can be both so constant and so widely distributed. Whether in cilia as doublets or in the mitotic spindle as single bundles, most microtubules studied have thirteen subunits of dimer protein making up their walls (fig. 26).

The shafts of all motile cilia are even more consistent. With few exceptions they display variations on the same theme. In some very high quality micrographs the standard set of tubules appear to have extra tubules (fig. 27A). Cilia have a peculiar ninefold symmetry of microtubules in double pairs (doublets), with two single tubules in the center. Looking somewhat like a telephone dial in cross section (see fig. 24A), the microtubule formation is called the "9 + 2 arrangement." Invariably 9 + 2 shafts (the axonemes) develop from charac-

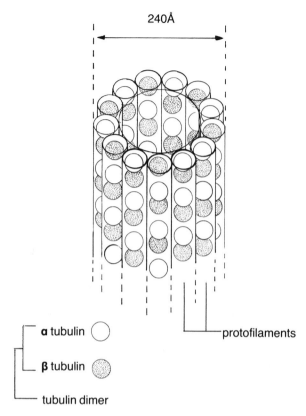

240Å

α tubulin ◯

β tubulin ◉

tubulin dimer

protofilaments

Fig. 26. Substructure of a microtubule (for example, from the mitotic spindle, axoneme, kinetosome, or axon of a neuron). α and β tubulin, each of molecular weight of about 50,000 dalton, make up a tubulin dimer. Thirteen subunit dimers make up the wall of the tubule. (Drawing by Christie Lyons.)

teristic bodies of the same width at the bases of the axonemes. These particles are called "centrioles" or "kinetosomes" ("basal bodies" in the older literature on the subject).

Centrioles display a 9 + 0 arrangement of microtubules. They measure 200–270 nm in diameter and from 175 to 700 nm in length. They are distinguished from kinetosomes only by the presence of the emerging 9 + 2 shaft or axoneme above the kinetosome, an intrinsically motile axial thread that is covered by the cell membrane. Kinetosomes begin their development as bodies identical to centrioles, but before and during ciliary axoneme development they become more elaborate (Wheatley, 1982). *Flagella*, at first, was simply the name given to moving "cell whips," cellular appendages related to motility. Over the past decade it has become abundantly apparent that there are two entirely different kinds of flagella. Those of bacteria have no relation whatever to those of eukaryotic cells. They rotate in a wheellike fashion and are composed of only a single filament.

"Flagella," long, undulating organelles often found in small numbers on eukaryotic cells, were revealed to be identical in transverse section to cilia (fig. 27B). Their axonemes also have the 9 + 2 arrangement of microtubules. We therefore retain a term, suggested in the 1930s, for 9 + 2 microtubule-based organelles (Corliss, 1979). This term is *undulipodia,* which refers to both cilia and 9 + 2 eukaryotic flagella. The term is necessary because cilia and flagella represent identical structures in evolution. To avoid confusion, it is better to restrict the term *flagella* to the entirely different structures found in prokaryotes.

Included in the list of eukaryotic structures composed of microtubules are the mitotic spindle (bundle of tubules), the asters of mitosis (rosettes of tu-

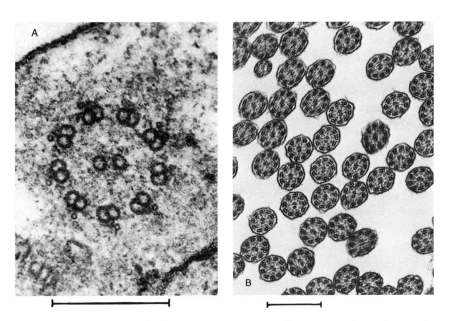

**Fig. 27.** A. Minimicrotubules in *Hexamita muris.* An axoneme from the protist *Hexamita muris* appears to have extra "minimicrotubules" (bar = 0.25 micrometers). (Courtesy of Dr. David Chase.) B. Axonemes of the protist *Trichonympha.* Axonemes, the 9 + 2 microtubular shafts of undulipodia, are identical in "cilia" (undulipodia) and eukaryotic "flagella" (also undulipodia) (bar = 0.5 micrometers). (Courtesy of Dr. David Chase.)

bules), suctorian tentacles, and mastigote axonemes (patterned alignments of tubules). (See table 7 for more examples.) Electron microscopic studies have confirmed the idea of early cytologists that these microtubule-based structures are related by common descent. What is in question is their nature. In particularly excellent micrographs, before the finished, patterned organelle appears (be it axopod, cilia, sperm tail, or haptoneme) one can discern its MTOC. These granular or fibrous bodies are the MTOCs hypothesized by Pickett-Heaps to be the central organizers for the structuring of microtubules.

The concept of MTOC was developed in order to explain the appearance of localities in the cell at which up to tens of thousands of microtubules organized to form distinct, perfectly patterned structures characteristic of the species, age, and genetic background of the particular cell in which they were found. Recently attention has focused on the source of pattern, the place in the cell at which the microtubules begin to organize. For example, in mouse embryos undergoing ciliogenesis (undulipodiogenesis) definite material can be seen adjacent to the centriole. This material, called "pericentriolar material," provides us with a clear example of an MTOC. Since specific antibodies can be produced against pericentriolar material and then used to stain it in electron microscope sections, at least part of the MTOC is proteinaceous (Calarco-Gillam et al., 1983). Professor Joan Steitz at Yale University has shown that the pericentriolar material contains RNA. This MTOC thus appears to be composed of ribonucleoprotein (RNP).

The number of MTOCs in a cell increases, presumably by replication. All the characteristic, asymmetrical, microtubule-based cell structures that develop from MTOCs, including the axons and dendrites of brain cells, are not only underlain by microtubules, but the microtubules themselves are composed of unmistakably related proteins: alpha and beta tubulin. Alpha and beta tubulin, with molecular weights of about 50,000, are extremely acidic proteins containing excess glutamic acid. Each is composed of about 450 amino acid residues.

Tubulins are conservative proteins found in all eukaryotes with microtubules that have been studied. Detailed differences in these microtubule proteins promise to provide a basis for understanding evolutionary relationships among protists (Adoutte, Chaisse, and Cance, 1984). Alpha and beta tubulins are clearly related to each other by evolution from a common ancestral gene (Little et al., 1984). Moreover, beta tubulin from vertebrate brain is closely related to beta tubulin from various protists—more closely related, in fact, than it is related to alpha tubulin from vertebrate brain (Little et al., 1984). It is estimated that undulipodia are composed of as many as 200 other proteins in addition to tubulin. The full evolutionary story requires far more research into these proteins and RNA.

**Table 7.** Some Examples of Structures Made of Microtubules

*Monera*
    none

*Protoctists*
    mitotic spindle fibers, macronuclear tubules
    kinetosomes, undulipodia
    oral membranelles, haptonemes
    cirri
    axostyles (parabasalids)
    sperm-tail axonemes
    suctorian tentacles
    axonemes (heliozoa, other actinopods)
    pharyngeal baskets (nassulid ciliates)

*Fungi*
    mitotic microtubules
    cytoplasmic tubules associated with nuclear migration

*Animals*
    mitotic spindles
    campaniform sensory receptors
    tactile spines
    developing lens of eye
    axons
    dendrites
    melanocyte processes
    asters, centrioles, kinetosomes, undulipodia
    sperm-tail axonemes
    auditory kinocilia
    olfactory sensory cilia
    gustatory sensory cilia

*Plants*
    mitotic spindle fibers
    sperm-tail axonemes
    cellulosic wall-orienting tubules

An extensive review of 9 + 0 centrioles in many different kinds of organisms has recently become available. Our conclusion from the hundreds of references in the primary literature is that "possession of a centriole endows the cell with the ability to form a cilium" (Wheatley, 1982). In spite of years of claims that centrioles are required for mitosis, there is still no evidence that the 9 + 0 centriole itself plays any role in mitotic cell division. In looking for principles to explain the role of nearly ubiquitous centrioles in eukaryotic cells, "One is often left simply turning out misconceptions and offering nothing in exchange. The centriole, to most cell biologists, is an enigma" (Wheatley, 1982).

The observations collected by Wheatley and others can be united with a single hypothesis: undulipodia (cilia and eukaryotic "flagella"), the complex, waving organelles of eukaryotes, always underlain in their development by centrioles, were once highly motile, free-living bacteria (probably those similar to gram-negative, helically coiled spirochetes; Margulis, 1981; Margulis, To, and Chase, 1981b). The centriole is a spirochetal remnant required for undulipodial development. There is indeed a mass of circumstantial evidence to support this idea. Definitive proof, not yet available, awaits comparison of the tubulin and other undulipodial proteins, as well as the kinetosomal RNA, with the proteins and RNA of carefully chosen spirochetes. In the meantime some observations are relevant.

Spirochetes tend to associate with and actually hook on to neighbor organisms as a method of acquiring constant nutrition. We envision centrioles as remnants of the attachment sites of these spirochetes. It is agreed that centrioles are absolutely required for morphogenesis of mature, functioning undulipodia. In the symbiotic view, spirochetes—autopoietic, replicating, and motile entities—entered hosts. In time the elements of the spirochetal autopoietic system were integrated with those of the host in a series of symbiotic steps similar to those that occurred in the evolution of plastids and mitochondria from free-living bacteria. This integration led to the evolution of mitosis and an associated adaptive radiation of protoctists. The ubiquity of the centriolar structure and its 240-angstrom microtubules composed of tubulin proteins betrays this common ancestry.

Spirochetes are symbiotic at many levels. They swarm in unison, swimming in synchronous waves inside the hindguts of termites; they attach to and propel larger microbes. Spirochetes have been seen to enter, swim inside, and divide inside other organisms. Many are anaerobic organisms at home in the intracellular environment. They thrive in mammals, for example, in the gums of healthy people or attached to the cells of rabbit testicles. If the common ancestry is related to the symbiotic penchant of spirochetes so prevalent in nature, the vast number of clearly related structures is more easily understood.

The phenomenon of teeming, swarming, symbiotic spirochetes invading and then living mutualistically in the first protists was a precondition for both mitosis and the related appearance of meiotic sex. The spirochetes, in addition to attaching to cells to form propelling undulipodia, were stripped of redundancy; many of their nutritional and biosynthetic needs had already been supplied by nucleocytoplasm and mitochondria. If we can be believed, the only function that needed to remain from those of the original spirochetal populations was replication of the entity responsible for undulipodial morphogenesis and movement. This is provided by the MTOC. The replicating spirochetal remnants became the MTOCs, the subvisible bases of the generation of characteristic microtubule-rich cell shapes. Replicating spirochetal remnants at the molecular level still persist to form the mitotic spindle and other microtubular structures. Alluding to the tendency of bacterial symbionts to fade into the cell of which they form a part, Professor David C. Smith of Oxford University compared the remains of endosymbiotic organelles to the smile of the fading Cheshire cat in *Alice in Wonderland* (Richmond and Smith, 1979).

MTOCs represent the streamlined essential remains of once-bustling populations of symbiotic spirochetes. Certain structures involved in mitosis—such as all MTOCs (which we expect will always be composed of ribonucleic acid and protein), and in particular the MTOCs called "kinetochores," dotlike structures on the chromosomes—are also taken to represent latter-day derivatives of these symbiotic bacteria. These organelles, responsible for the attachment of chromatin to the spindle and subsequent separation and movement of chromosomes to opposite cellular poles prior to cell division, are interpreted by us to be pared-down spirochetes whose essential capability for replication and generating movement have withstood the test of time in a living environment. To understand the relationship between bacterial symbiosis and the replicating, electron-dense material that comprises MTOCs, which play such a fundamental part in meiotic sex, it is necessary to think about how spirochetes could have become MTOCs, a transition crucial to the origin of mitosis. The idea of the spirochetal origin of undulipodia has been discussed, supported, and criticized in detail (To, 1978; Margulis, Chase, and To, 1979; Dodson, 1979; Margulis, 1981; Wheatley, 1982), and even dismissed (Roos, 1984); it has certainly not been firmly established.

The spirochete hypothesis, like the hypothesis of bacterial ancestry for plastids and mitochondria, asserts that motile bacteria became, with time, undulipodia. However, the level of integration of the spirochete precursors to undulipodia must far transcend that of mitochondria and plastids. It is as if the Cheshire cat has had a litter of kittens that are playing everywhere, yet more enigmatic and faded than their ancestor.

We have speculated that spirochetal remnants were crucial to the origin and evolution of sex for a variety of reasons, among which are the following. It is common for symbionts to undergo reorganization of their genetic components; spirochetes undergo transformations living inside larger cells; spirochetes are remarkable among bacteria for their rapid motility. Microbes are known to give up metabolic pathways or even structural parts of themselves after living endosymbiotically within other cells. *Aeromonas hydrophila* bacteria, symbiotic with the animal hydra, lose some abilities to metabolize carbon compounds (Thorington, 1978): their free-living cousins are able to metabolize a wider range of nutrients. Still more impressive is *Mesodinium*, a protist whose algal (cryptomonad) symbionts break up into organelles that reproduce at different rates (Taylor, 1983). This reconstitutive genetic process is postulated to have been much more profound in symbiotic spirochetes because their motility apparatus was so useful to their host organisms. The ubiquity of intracellular motility in eukaryotes and its total absence in prokaryotes we attribute to the symbiotic acquisition of spirochetes in the ancestral populations of all mitotic organisms. We believe the intracellular spirochete bodies were used to form propellant undulipodia and the microtubular motility apparatus of eukaryotic cell division itself, both mitosis and meiosis. Today the *Hollandina* and *Treponema* spirochetes of the protist *Mixotricha paradoxa* are so entrenched in their symbiosis that, though they are clearly recognized as spirochetes, no one has ever succeeded in growing them apart from their host. The same is true of the *Treponema* symbiont of the protist *Pyrsonympha*, which has evolved special attachment apparatus to keep it "plugged in" to its host. The dividing spirochetes apparently remain attached to their hosts. Indeed, if the motile, tubulinlike proteins of these spirochetes were exploited and integrated into the host genome, and if their structure were more regular, we would be forced to consider them undulipodia (flagella or cilia).

Some bacterial genes have been lost and others transferred to the nucleus or to other organelles. We believe that the fusion of motile spirochetes with other members of the bacterial community—primarily the mitochondria—occurred at virtually every level: spirochete genes were lost, transplanted, and duplicated; spirochete ribosomes fused with host ribosomes to form the larger "80S" ribosome of eukaryotes; cell motility systems and spirochete nucleic acids became incorporated into host nuclei to form emergent entities—chromosomes displaying coiling cycles and great quantities of excess DNA. The first protists began as loose confederacies and communities, but they emerged as rigid bureaucracies and distinctive regimes. A summary of the status of the evidence for the origin of undulipodia from spirochetes is presented in tables 8 and 9. The probable answers to questions generated by the spirochete hypothesis are in table 8 and undulipodia are compared to spirochetes in table 9. For a summary of the status of the symbiotic theory of the origin of organelles, see Margulis and Stolz (1985).

## ORIGIN OF THE NUCLEAR MEMBRANE

Every eukaryotic cell contains a nucleus of DNA packaged with protein. In nearly all animals and plants, and in many protoctists, the DNA protein complexes are arranged into *chromosomes:* countable, stainable, and clearly distinguishable bodies easily seen with a standard light microscope. By definition, the material comprising chromosomes is chromatin. Since sectioning of cells for electron microscopy usually cuts through chromosomes, making their morphology and number difficult to assess, the method of choice for studying chromosomes is light microscopy.

Many protoctists and fungi lack clearly distinguishable and countable chromosomes; instead, inside their nuclei one finds fuzzy threads interpreted to be DNA and protein homologous to the visible chromosomes of most eukaryotes and thought to be comprised of chromatin. Often these sorts of nuclei of protoctists and fungi fail to take on standard stains for nucleic acid such as Feulgen and acetic acid orcein. It is highly likely that the poorly visible fungal and protoctist chromosomes, composed of so-called uncondensed chromatin, have a somewhat different protein composition from those of protoctists, animals, and plants that during mitosis form "condensed chromatin" and display sharply visible chromosomes. The most likely differences between the nearly invisible chromatin of organisms such as *Enteromorpha* (a zygomycotous fungus) and *Achlya* or *Saprolegnia* (oomycotous protoctists) and most plant and animal chromosomes are the absence of certain histone proteins (such as histone 1) and a somewhat different organization of the nucleosomes, the small bodies aligned along chromatin visible with the electron microscope (Silver, 1984). In certain amoebomastigotes (such as *Paratetramitus;* Margulis and McKhann, n.d.) and other protoctists, such as *Nanochlorum eucaryotum* (Zahn, 1984) and *Pelomyxa* (Whatley and Chapman-Andresen, n.d.), no clearly visible chromosomes or DNA are seen with any technique at any stage of the life cycle. Although it is assumed that the amorphous fuzz inside the nuclei in micrographs of these organisms represents chromatin DNA and protein, the details of the organization of intranuclear materials remain unstudied.

The DNA of the chromosomes—which vary in numbers from species to species from two to more than a thousand—is surrounded by RNA and several proteins. The whole center of the cell is outlined by a single-layered membrane, often with pores, that separates the nuclear genetic material from the rest of the cell, the cytoplasm. How the nucleus evolved as an organelle in the ancestors of the phenomenally successful eukaryotes is a subject of debate.

**Table 8.** Probable Answers to Questions Generated by the Spirochete Hypothesis

| | |
|---|---|
| What was the free-living form of protoundulipodia? | Anaerobic or aerotolerant motile prokaryotes, spirochetes, or spiroplasms containing tubulin microtubules. |
| In what host cell did the protoundulipodia first become established? | In heterotrophic cells with nuclear membranes, some containing mitochondria. |
| What environmental agent selected for and maintained the symbiosis? | Scarcity of food: protists made motile by acquisition of spirochetes sought nutrients more efficiently. |
| When did these symbioses become established? | In the Proterozoic Eon, after the transition to the oxidizing atmosphere and as a prerequisite to the evolution of mitosis. |
| Is this symbiosis obligate? | Yes, in all mitotic organisms, because of the utilization of "motility proteins" from the free-living spirochete for mitosis and other intracellular motility phenomena. |
| Which traits of free-living prokaryotic cells do the MTOCs and undulipodia retain? | Nucleic acid and very limited protein synthesis. Kinetosomal RNA resides in the lumen of the kinetosome (Brinkley and Stubblefield, 1970; Dippell, 1976). New RNA synthesis accompanies new kinetosome production (Younger et al., 1972) and that kinetosomal RNA functions in MTOC; genes for kinetosomal morphogenesis on single linkage group may be RNA (Dutcher, 1984). No DNA in kinetosome (Randall and Disbrey, 1965; Smith-Sonneborn and Plaut, 1969); former evidence of "kinetosome DNA" can be explained as contamination by mitochondrial DNA (Younger et al., 1972) and by RNA (Hartman, Puma, and Gurney, 1974). |
| What did the protoundulipodia lose after becoming symbiotic? | Cell walls and plasma membranes; most biosynthetic functions, including synthesis of DNA and ribosomal proteins, were relegated to the nuclear gene-controlled nucleocytoplasmic system; only ribonucleoprotein synthesis was retained at site of undulipodia development. |
| What changes the ratio between the number of the original host genomes (now nucleocytoplasm) to the number of MTOCs? | Tissue differentiation (Sorokin, 1968), life-cycle stages (Crocker and Dirksen, 1966), and speciation (especially in ciliates, Corliss, 1979); subjection of eukaryotic cells to low concentration of vitamin E and high concentration of oxygen (Hess and Menzel, 1967); treatment with Colcemid (Stubblefield and Brinkley, 1966); treatment with pargyline, a monoaminooxidase inhibitor (Monaud and Pappas, 1968); and pituitary hormones (Dustin, 1978). |

108

| How can the loss of the symbiont from the host organism be induced? | Loss is presumably lethal in all mitotic organisms. Cleveland (1956) treated developing hypermastigote gametes with low concentrations of oxygen; this caused preferential destruction of the chromosomal system, producing a nucleated cell still able to divide. |
|---|---|
| How much may 9 + 2 homologues dedifferentiate? | To below the limit of resolution of the electron microscope (Schuster, 1963; Stubblefield and Brinkley, 1966; Crocker and Dirksen, 1966); in many protoctists, plants, and fungi, MTOCs apparently are normally dedifferentiated to that level, presumably ribonucleoprotein only. |
| How are offspring MTOCs produced? | Usually from parent MTOCs; for example, new kinetosomes form at right angles to the parent in a rather complicated sequence that has many variations (Mizukami and Gall, 1966; Crocker and Dirksen, 1966; and fig. 49). Centrioles and associated nuclear organelles that duplicate in mitotic organisms before becoming visible structures must have undergone nucleic-acid replication. |
| What organisms evolved from spirochete-bearing aerobes with mitochondria? | All mitotic eukaryotes. |
| What do the genomes of the original spirochete symbionts code for? | Not known; not tubulin, possibly RNP such as 6S or some RNP of the 80S ribosomes. Membrane-associated RNA synthesis is probably related to kinetosome replication (Dippell, 1976; Heidemann, Sander, and Kirschner, 1977; Hartman, Puma, and Gurney, 1974). The tubulins, microtubulin associated proteins (MAPs), and other undulipodial proteins should be homologous with those in appropriate spirochetes (Margulis, Chase, and To, 1979; Fracek, 1984; Obar, 1985). |
| What is the evidence for genetic autonomy of kinetosomes? | Inheritance of stentor undulipodia independent of nucleus (Tartar, 1961); inheritance of manipulated cortical patches in paramecium independently of mitochondria and nucleus (Beisson and Sonneborn, 1965). Growth and development of kinetids with their undulipodiated bands in hypermastigote gametes in the total absence of the nucleus (Cleveland, 1956). |
| How are MTOCs transmitted from parent generation to offspring generation | Centrioles carried in sperm in many animals (E. B. Wilson, 1925; Turner, 1968; Wheatley, 1982). Unknown. |

(*Continued*)

**Table 8.** *(Continued)*

| | |
|---|---|
| through complex eukaryotic life cycles? | |
| Why are most undulipodial functions under nuclear genetic control? | Mitosis requires very close coordination between nuclear chromatin and the MTOCs responsible for its segregation; selection against redundancy and for mitosis led to transfer of indispensible spirochete genes to nucleus. |
| What original symbiont functions have been relegated to the host nucleus? | Nuclear genes control the development of the inner two microtubules in *Chlamydomonas* axonemes (Randall et al., 1964). Tubulin and MAPs are nuclear gene products. Nearly all original symbiont functions will probably be found to be under direct nuclear control. |
| Why do the numbers of centrioles and undulipodia per cell vary, yet the unit sizes of microtubules, kinetosomes, and undulipodia remain constant? | At first, they varied as a consequence of the fact that the symbiont/host ratio is seldom 1/1; as control evolved, they varied because of differentiation for motility and their role in mitosis. Size is determined by the original spirochete symbiont genome; like the diameter of prokaryotic cells, the size of individuals in a population varies very little. |

Whatever the answer, the nucleus and its membrane clearly came before mitosis. No mitotic organism lacks a nucleus, whereas several types of protists, such as the anaerobic *Pelomyxa palustris*, have membrane-bounded nuclei but lack mitosis. Thus we can separate the origin of the first nuclear membrane from that of mitosis and its microtubular apparatus. Although the three schemes we recognize for the origin of the nuclear membrane are hardly even hypotheses, we cannot continue the narrative of the origin of mitosis without assuming the membrane's existence. For clarity, let us call theories of the origin of nuclear membrane the "untangled DNA" hypothesis, the "oxygen protection" hypothesis, and the "foreign nuclear symbiont origin" hypothesis.

Because eukaryotes have so much more genetic material than prokaryotes, and are larger, the untangled DNA hypothesis suggests that the earliest bacterial-prey complexes on their way to becoming the first eukaryotic cells mutated to develop an internal DNA-enclosing membrane, elaborations of the well-known mesosomal membranes of bacteria. Once the membranes hypertrophied they could distribute the newly synthesized DNA more efficiently inside the enlarging cell. Each elaboration in the partitioning of genetic material on membrane subsequent to replication was selected for when it proved able to distribute more genes more evenly to offspring cells.

**Table 9.** Comparison of Spirochetes and Undulipodia

| CHARACTERISTICS | UNDULIPODIA | SPIROCHETES |
|---|---|---|
| Size (μ = micrometer) | $0.25 \times (1\text{–}15)\mu$ | $(0.09\text{–}3.0) \times (1\text{–}100)\mu$ |
| Flagella (axial filaments) | – | + |
| Microtubules seen by electron microscopy | universal | not universal; observed in *Pillotina, Diplocalyx,* and some small termite spirochetes |
| Organization of microtubules | 9 + 2 array | longitudinal alignment; no 9 + 2 array |
| Independent helical motility | + | + |
| Motility ATP sensitive | + | – |
| Motility proton-pump sensitive | ? | + |
| DNA | – | + |
| RNA | + | + |
| Proteins | $10^2\text{–}10^3$ | $> 10^3$ |
| Tubulin reacts with antibody to brain tubulin | + | + (*Pillotina, Hollandina, Diplocalyx, Spirochaeta bajacaliforniensis* BA2 and BA4, *Spirochaeta halophila;* see figure below) |
| Molecular weight (K = 1,000 daltons) | 55K (α, β) | 67K (S₁) |
| Peptide analysis* | + | similar to brain tubulin |
| Entire amino acid sequence | + (tubulin α) | not available |
| Purification by warm-cold cycling* | + | + |

*Key:* S₁—spirochete protein from *Spirochaeta bajacaliforniensis;* α, β—tubulins from mammalian brain. (Fracek and Stolz, 1985)
*Obar, 1985.

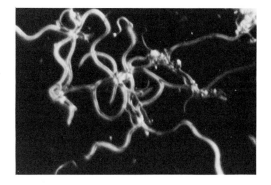

Treated with an antibody raised against mammalian brain tubulin protein, these large spirochetes from the Pillotaceae family (see also fig. 37A and 37C) show a positive reaction, indicating the presence of tubulin-like protein inside them. Microtubules in transverse section from an unidentified spirochete from *Kalotermes schwartzi,* a South Florida termite.

The oxygen protection hypothesis holds that a membrane-bounded nucleus came into being in order to protect vulnerable genetic material from the combustive properties of oxygen, a reactive gas that still poisons many cells and that was produced in vast quantities around the time of the origin of the first protists by burgeoning global populations of photosynthetic bacteria. Several classes of lipids, polyunsaturated fatty acids, and steroids found regularly in high quantity in eukaryotic membranes are much less frequent or entirely absent in bacterial cells. It is thus theorized that the cellular use of steroids—a group of lipids that requires oxygen for its manufacture—may have been important in keeping the gas away from what was to become the cell's nucleus. Free oxygen at very low concentrations chemically reacts with the steroid precursor (squalene) and is used in the cyclization of the squalene chain to steroid precursor molecules such as lanosterol (in plants), cycloartenol (in fungi), and ergosterol (in animals). In this way steroids, first made because they were chemically responding to oxygen, were later retained for protection and other functions in membranes (for example, they confer great flexibility and differentially permit passage of other lipids through membrane). By making such steroid membranes, a process involving both nucleocytoplasmic and protomitochondrial metabolites, the new cells could avert oxygen damage and put the oxygen-utilizing synthesis to good use. Steroids, active in making and fusing flexible lipid-protein membranes, were first used in membranes that wrapped nuclear DNA. They were also employed in the development of vacuoles, in endoplasmic reticula, and in outer mitochondrial and other typically eukaryotic endocellular membranes.

The third hypothesis (which is not part of the SET) suggests that the entire nucleus is a remnant of a symbiotic nucleated microbe that lost its cytoplasm (Hartman, 1975). This idea simply pushes back the problem of the origin of the nucleus to its origin in the unidentified nucleated symbiont. (The symbiont-nucleus hypothesis is comparable, theoretically, to the explanation of the origin of life by invoking "directed panspermia": saying that it came from an extraterrestrial spacecraft is similar since it effectively defers rather than directly answers a question of origin; Crick, 1981.) Proponents of the symbiont nucleus hypothesis have not detailed what became of all of the original host microbe's DNA. Another problem with the symbiont-nucleus theory is that ribosomes—the organelles in the eukaryotic cytoplasm upon which protein synthesis takes place—are now entirely integrated in their interaction with the nucleus. Nuclear DNA provides the complement for messenger RNA, which then leaves the nucleus, attaches to ribosomes, and serves to provide the information for protein synthesis. It seems most likely that the eukaryotic cytoplasm first evolved autogenously. The nucleus evolved by partitioning off nuclear material inside the membrane of a single protein-synthetic unit. Furthermore, we know of no precedent of a eukaryotic symbiotic microbe that left only its nucleus inside another cell and that in so doing has destroyed the entirety of the host cell's DNA.

In explaining the evolution of the nucleus as a eukaryotic DNA segregation device by way of the membrane, perhaps both of the first two hypotheses are correct: the membrane that neatly segregates replicating DNA also protects nuclear contents from oxygen damage. In any case it was apparently to the great evolutionary advantage of complex organisms to package their DNA into discrete, movable units and wrap them in membrane. Cell membranes, in bacteria and eukaryotes alike, probably begin synthesizing at sites of attachment of lipid to DNA and protein. The complex steroid-protein membranes of eukaryotes are likely to have been products of dual genomes. These membranes, perhaps originally involved in segregating nuclear DNA during the period of its synthesis, preadapted eukaryotic cells to a life of engulfing. Field and laboratory observations alike document the predilection of eukaryotic cells to engulf and "examine" many things: prey bacteria, particles of sediment, glass, and today even plastic beads. This tendency to engulf and examine is another aspect of the preadaptation of eukaryotes for what, in the end, became the meiotic sexual cycle.

# 7 · THE ORIGIN OF CHROMOSOMES

## Packaging of Chromatin

### THE USE OF MOLECULAR GENETIC INFORMATION

In this chapter we will look at the specific structural differences of DNA organization between bacteria and meiotic eukaryotes. By examining unique organisms such as the dinomastigotes, which exhibit structural features of both, it will be possible to trace the path from direct bacterial cell division to mitosis. The expanding armamentarium of tools developed by molecular biologists permit us far more precision in our evolutionary analyses than ever before.

There are many differences between the organization of the prokaryotic genome, or genophore, and the genomic organization of eukaryotes. To understand the emergence of meiotic sexuality, the properties of its prerequisite, mitosis, must be kept in mind. The comparison of standard mitosis and meiosis in plants and animals with cell division in bacteria sheds light on the complex evolution of genetic information and its packaging. Variants on chromosome organization in protoctists represent some of the meanders that have taken place as modes of distribution of genes to offspring generations have evolved. We shall begin here with the salient features of eukaryotic genome organization into chromatin, as well as the deployment of the chromatin to offspring cells in mitosis. Our examples are generalizations drawn from the best studies.

The emergence, in the mid-1970s, of direct methods of studying nucleic acid sequences and their processing has led to the accumulation of vast quantities of data, the interpretation of which in many cases is obscure. We shall emphasize here the well-established observations and data, ignoring for the most part the interpretations proffered by the many biochemical investigators themselves.

One of the clear results of eukaryotic molecular genetics research is the revelation of an astounding number of repeated sequences of DNA that have no direct role in the coding of proteins. A second observation indicates that methods for recombining nucleic acids from different sources and for rearranging, detecting, and copying nucleotide sequences are rampant both in prokaryotes and in the nuclei and organelles of eukaryotes. Perhaps the most serious shock, revealed by hundreds of investigations, is the general inconstancy of nucleic acid sequences. DNA and corresponding RNA sequences generally show far more variation than was anticipated by the conservative nature of the morphology and protein chemistry of life. Whereas proteins, especially those structural proteins with functions that date back hundreds of millions of years, show remarkable constancy from organism to distantly related organism, variations in nucleic acid sequences abound. Cytochrome c sequences can be traced from photosynthetic bacteria to the mitochondria of wheat. Certain histone proteins (H3 and H4) of cows, peas, and even fungi (yeast) are nearly identical—they differ in only a few amino acid derivatives. Tissues in the organs of individuals, members of the same species and related species, even organelles from the same individual, show remarkable identities in their major metabolic pathways. Yet, while the enzymes that catalyze these metabolic reactions are often very similar, there exist vast differences in the number and quality of the sequences of nucleotides comprising the cellular nucleic acids.

The excesses of nucleotide base pair sequences that do not function in the determination of amino acid sequences in protein have led to concepts of "selfish DNA," ideas that "extra" DNA is in cells simply because DNA possesses the ability to replicate and maintain itself, much as offices may remain full of bureaucrats long after they have ceased to serve a useful function (Doolittle and Sapienza, 1980). We doubt that "selfishness" is the primary explanation for the excess DNA in eukaryotes. Although the paucity of understanding of the possible noncoding functions of nucleic acids in development leads us to restraint in interpreting the molecular biological data directly, we note that chromosomes are no more just large genophores than euglenids are little plants. The enormous lengths of DNA packaged into chromosomes become some 8,000 times smaller after packaging into the eukaryotic genome (Strogatz, 1983). Evidence from studies on ciliates discussed below indicates that at least some "junk" eukaryotic DNA may be directly involved in packing problems. Excess DNA may be present for organizational tasks that have no prokaryotic counterparts:

In this book those molecular biological observations most relevant to an understanding of the evolution of sexuality are brought into play; biochemical information is used only in a supplementary fashion. Eukaryotic chromosomes and the sexual processes they engage in are emphasized. Although we suggest possible molecular biological consequences of the evolutionary odyssey we feel to have been most likely, our narrative remains on the cell and organismal level. Our discussion of biochemical observations is purposely limited to those that are directly useful in interpreting biological processes. We will indicate those aspects of the scenario that are clearly speculative.

## CHROMOSOMES AND CHROMATIN

Except in those protoctists that seem to lack them, chromosomes are composed of nucleoprotein organized into nucleosomes. DNA, in other words, is wrapped around sets of histone and nonhistone proteins in a highly regular fashion to form a 100-angstrom-wide fibril. The fact that this unit chromatin fibril is not present in some protoctists even though they have nuclear membranes shows that these eukaryotic characteristics—namely, nuclear membranes and chromatin fibrils—were initially separate. They evolved at different rates: the more universally distributed nuclear membrane appeared first, before the chromatin unit fibril.

Eukaryotes, unlike prokaryotes, synthesize their cellular DNA in a punctuated manner. Prokaryotes continuously synthesize DNA throughout their cell cycle. In eukaryotes, only during a portion of the cell cycle (the time called the "S" or "synthesis period") does DNA actually undergo synthesis to produce a complementary DNA. During this period chromatin tends to be difficult to stain and visualize clearly. After the S period there is a temporary lapse in active synthesis of cell DNA and protein that is called the "G2" or "growth 2 period." After G2, chromosomes form from chromatin and become visible. At the end of the G2 period, prior to mitotic cell division, chromatin dramatically "condenses" to form chromosomes. The long, thin chromatin strands become short and thick and can be easily stained. The chromosomes take on a number and a form typical of the species of organism in question (Cleveland, 1953). This process, called the "chromosome coiling cycle" or the "condensation" of chromatin, has been seen in live and stained cells since the end of the nineteenth century (E. B. Wilson, 1925; Cleveland, 1949). Great lengths of DNA are packaged into tiny volumes. (In people, for example, the total DNA in all the chromosomes of all cells would reach out more than a billion kilometers if stretched end-to-end.) As Strogatz (1983) says, "A thin but stiff cable is somehow wrapped, looped and folded to fit within a container whose linear dimensions are several times smaller. The packaging of DNA is, without exaggeration, an engineering feat of staggering proportions" (p. 602).

Whether the wrapping, looping, and folding is directly related to the presence of certain "motile proteins" attached to chromatin is not known. It has recently become clear, however, that chromatin is intimately associated with some familiar proteins of eukaryotic cell motility: the fibrous protein actin and the globular, ATP-degrading protein myosin. These have been localized in chromatin in several protists by immunofluorescence and other methods (Omodeo, Capanna, and Pallini, 1982; Soyer-Gobillard, 1982). We suspect that proteins homologous to those responsible for muscle motility in animals—the actin and myosin that form actinomyosin filaments or muscles—are probably involved in the "condensation" and the "decondensation" properties of chromatin.

In almost all eukaryotes the DNA includes often highly repeated DNA sequences interspersed with the unique sequence DNA. Only the latter is transcribed into messenger RNA and determines the corresponding order of amino acids in enzymatic and structural proteins. The functions, if any, of the repeated DNA sequences have not been established. While some of the dispersed repeated sequences are about 100–300 base pairs in length, the single-copy sequences are more like 1,000–3,000 base pairs long. But repeated sequences up to thousands of base pairs long have also been detected, as have inverted sequences. An inverted sequence has a string of recognizable base pairs that extends from the 3-prime (hydrogen-oxygen) end of the DNA rather than the 5-prime (hydrogen) end. A class of excess DNA known as "pseudogenes" can also be recognized. These are nucleotide base pair sequences that are clearly related to functioning gene DNA sequences but in which certain base pairs are missing or so severely changed that they cannot be transcribed into functional proteins. There is no consensus concerning the explanation of this familiar, yet apparently nonfunctional, DNA. Any of these classes of repeat DNA may bind to relevant proteins and form part of the modulated structure of living chromosomes. That is, some of the "excess DNA" may be part of the solution to the engineering problem of packaging DNA alluded to above. Whatever its ultimate explanation, however, such DNAs have been found in the chromatin of all animals and plants in which they have been sought.

Chromatin is composed of more protein than DNA; by weight the ratio of both histone and nonhistone chromosomal protein to DNA is over 60 percent. By number the ratio of total histone proteins to DNA is approximately one-to-one and stays that way; unlike all other proteins, which vary enormously depending on cell type, age, response to hormones, and other factors, histones (1, 2A, 2B, 3, and 4) and DNA are synthesized in synchrony during the S period.

We have seen how chromatin, fundamentally very long, thin threads, is organized into *nucleosomes*, interspersed beads along the DNA. Nucleosomes may have evolved from similar, but smaller structures in *Thermoplasma*-like hosts (Searcy, 1982). Thermoplasma are extant bacteria that Searcy has argued are most closely related to the original nucleocytoplasm of eukaryotes (Searcy and Stein, 1980). Nucleosomes, which may be ordered in a zigzag pattern to facilitate the exposure of locally unpackaged DNA, are oblate spheroids, $110 \times 110 \times 550$ angstroms, seen clearly in electron micrographs of chromatin. Nucleosomes are eight-part units (octamers), composed of four highly conserved histones (H2A, H2B, H3, and H4), each taken twice. Some 145 base pairs of DNA are wrapped around the exterior of these nucleosomes (fig. 28). Linker segments of DNA of varying lengths, bound by a more variable histone protein, called H1, occur between the nucleosomes. Histone proteins stabilize DNA from degradation by heat and protect it from breakdown by proteolytic enzymes. The histone wrapping around the DNA that comprises the nucleosome structure seems to prevent messenger RNA transcription. Apparently RNA-transcribing enzymes cannot attach to nucleosomal DNA; rather, they bind between nucleosomes or to unwound nucleosomal DNA. Histones, in the sense that they prevent direct interaction of DNA with polymerases, inhibit RNA synthesis. The nonhistone chromosomal proteins vary from organism to organism, tissue to tissue, and at different stages in the life cycle. Nonhistone proteins (NHP) include enzymes that have polymerase, ligase, endonuclease, kinase, and other activities. The ratios of nonhistone proteins to DNA vary from 0.2 to 0.8 (NHP:DNA).

Histones are relatively small proteins. H4, the smallest, has a molecular weight of about 10,000, whereas the molecular weight of the largest, H1, is 22,500. All histones are rich in positively charged amino acid residues; called "basic proteins," they take up positive hydrogen ions from solution because of their $NH_3$ groups. They have many arginine and lysine amino acids and lack tryptophan. The positive charges of the histones ($NH_4^+$) bind to the negative charges of the DNA ($PO_4^{3-}$) to form the beaded structure. It is inferred that histone structure must be conserved in order for cells to function and that any mutations in histones have tended to be lethal. The protoctist ancestor of cows, yeast, and peas, organisms separated by more than 500 million years of divergent evolution, already had nucleosomes composed of DNA-wrapped histone. The histone-DNA complex is so unchanging that today one can make histone-DNA complexes between organisms as distantly related as yeast (*Saccharomyces cerevisiae*) and calves, and they are just as tightly binding as histone-DNA complexes of yeast with yeast or calf with calf.

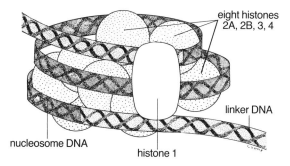

eight histones
2A, 2B, 3, 4

linker DNA

nucleosome DNA

histone 1

**Fig. 28.** Diagram of a nucleosome. The DNA double helix is wound around eight histones (two each of histones 2A, 2B, 3, and 4). Histone 1 (in front) binds to the outside of the nucleosome DNA and to the linker DNA. (Drawing by Christie Lyons.)

Although viral and bacterial DNA can, in the test tube, form complexes with histones isolated from eukaryotes, in nature the viruses and prokaryotes entirely lack chromatin made of nucleosomes and linkers. No changes in DNA are required to allow it to bind to histones; rather, histones and chromatin organization evolved in protists and never appeared in the bacterial world. Given that histone of peas and cows differ in only two amino acid substitutions, only one protein is known that is more conserved. It has an absolutely equivalent amino acid sequence in all organisms, all cells, and all tissues studied so far. Present in all organisms in which it has been sought, this small protein, appropriately called "ubiquitin," binds to DNA (Gold-knopf, 1978). In eukaryotes, when the genome is not being transcribed (that is, when messenger RNA is not being formed from a DNA complement), ubiquitin can be found as a part of histone 2A, one of the nucleosome proteins. Perhaps it is indispensable for DNA conformation in transcription or packaging in all organisms. At any rate, ubiquitin is the only clear link between the nucleoid of prokaryotes, which contains fewer than a few percent protein and none organized into chromatin, and the nuclei of plant and animal cells, which are always made of chromatin, composed of more than 50 percent protein.

Whatever the details, an evolutionary problem of great magnitude remains: what is the origin of the chromatin and chromosomal proteins? If there were no variation in the two extremes—the nucleoids of prokaryotes and the nuclear chromatin of eukaryotes—there would be no way to reconstruct genome evolution. The evolution of nuclear organization of eukaryotes could not have been traced in plants or animals because these organisms show no significant variation in these fundamental eukaryotic properties.

Fortunately, however, several groups of protoctists show dramatic variations on the chromatin theme. It is by their study that one can construct a plausible scenario for the evolution of chromatin. The major conclusions are twofold: first, that chromatin organization was already stabilized when the plant lineage diverged from the animal one; second, that chromatin organization evolved in common ancestors of animals, plants, and eukaryotic microorganisms and their descendants (protoctists). In order to understand the evolution of meiotic sex one must study the emergence of chromosome organization in concert with the life-styles of the relevant protoctists.

We shall see in the next chapter how chromatin organization, the folding and packaging of DNA, first evolved independently of chromosome movement. Well-packaged chromosomes later interacted with the microtubule system and its microtubule organizing centers, which provide the basis for the intracellular motility of chromosomes in mitotic and meiotic cell divisions. At the onset of the evolution of mitosis, chromosome organization and chromosome movement were separate phenomena, evolving at separate rates in different protoctist lineages. Vestiges of the evolutionary process remain in extant protoctists. With time, the microtubule organizing centers and the chromatin system stabilized and coevolved in a coordinated fashion. This occurred before the common ancestor of yeast, cows, and peas appeared, probably over 700 but certainly prior to 500 million years ago.

## DIDACTIC DINOMASTIGOTES

An odd group of whirling marine protoctists, of which there are over forty-two hundred species, provide the most insight into the evolution of chromatin in eukaryotes (Taylor, 1980; Soyer-Gobillard, 1982). Most of these eukaryotic microbes are single-celled, although some stacked colonial forms are known. Dinomastigotes lack nucleosomes and thus their chromatin organization is radically different from that of all other eukaryotes. The dinomastigotes have chromosomes, but strange ones numbering up to hundreds per nucleus, and they undergo a peculiar sort of mitosis. By comparing the DNA organization of entirely nonmitotic amoebae, *Pelomyxa*, with that of dinomastigotes, one can trace the path of eukaryotes up to the appearance of any chromatin.

*Pelomyxa palustris*, the nonmitotic amoeba, is a living relic of the life of the Proterozoic Eon subsequent to the evolution of nuclear membranes but earlier than the evolution of mitosis. Presumably pelomyxids and other eukaryotes, many of which probably formed resistant organic cysts (Vidal,

1984), were eventually out-competed by more efficient mitotic organisms. Some late-Proterozoic cysts may also be due to early dinomastigotes (Taylor, 1983). Whatever groups of organisms left their marks in the fossil record had hard-walled cysts. Whether green algae, chrysophytes, amoebae, ciliates, dinomastigotes, or other, the level of complexity, size, and appearance leads everyone who has seen them to conclude that many late-Proterozoic fossils were members of the kingdom Protoctista.

The existence of pelomyxids, entamoebae (Fahey et al., 1984), and several species of dinomastigotes that lack conventional mitotic spindles and nucleosomes indicates that the nuclear membrane evolved before the appearance of either mitotic spindle or chromatin. The chromatin organization and composition of dinomastigotes are so distinctive that all investigators studying them believe them to be products of evolution independent of the rest of the eukaryotes.

In most species of dinomastigotes structures shaped like bent rods composed of long, finely looped threads remain attached to the inner nuclear membrane. These threads—like bacterial DNA—are 25-angstrom-wide fibrils. They have been identified as the bearers of dinomastigote DNA. Even though there is a distinct period of DNA synthesis (the S period), the dinomastigote chromosomes do not decondense and recondense like the chromosomes of most other eukaryotes. They remain condensed, fibrous, and banded throughout the cell cycle. Although a small amount of positively charged protein, less than 0.2 percent, has been found associated with these peculiar chromosomes, the numbered histones of eukaryotes (1, 2A, 2B, 3, and 4) are unknown from dinomastigote nuclei.

Protein homologous to muscle actin has been found in the axis of the dense dinomastigote chromosomes. It is possible that actin comprises the central core of certain peculiar 90-angstrom-diameter fibers around which two helices of DNA are wrapped. Not only does this tight structure protect dinomastigote DNA from attack by DNAase enzymes, but dinomastigotes seem to be able to pack immense quantities of DNA around them. Human haploid sperm, themselves masters of DNA packaging, contain 3,300 micrograms of DNA. *Gymnodinium nelsoni*, a dinomastigote, contains 143,000 micrograms of DNA! Up to 60 percent of this DNA consists of repetitive sequences, again suggesting some relationship between packaging and "excess" DNA.

Instead of the normal thymines, dinomastigote DNA may have up to 70 percent substitution by another nucleotide base not generally found in nucleic acids of cells: hydroxymethyl uracil. The DNA unit fibril of dinomastigote chromatin, only 25 angstroms wide, is only about one-fourth the width of histone-bound, normal eukaryote unit fibril. During nuclear division the tight skeins of DNA in some dinomastigotes such as *Prorocentrum* are attached directly to nuclear membrane. No kinetochores, the structured attachment sites of chromosomes to microtubules found in most eukaryotes during cell mitosis, appear. The absence of histones, nucleosomes, and kinetochores, and the direct attachment of DNA to membrane are features that relate dinomastigotes to prokaryotes. On the other hand, the presence of mitochondria, undulipodia, and nuclear membranes argues for the unambiguous classification of the dinomastigotes with the eukaryotes.

The peculiarities of dinomastigote nuclear organization substantiate two generalizations: that mitosis as a process evolved in the protoctists and that the various components of the mitotic system (nuclear membrane, chromatin protein packaging, and MTOC mobility systems) evolved at different rates and under different selection pressures at different times. We have here reviewed the appearance of chromatin itself. In the next chapter we shall look at the evolution of the microtubule organizing centers and the mitotic spindles they produce. We shall see that the evolution of an entirely functioning mitotic system was an absolute prerequisite for the evolution of meiotic sexuality.

## NEITHER MITOCHONDRIA NOR MICROTUBULES

*Entamoeba*, a genus of parasitic protoctists that grow in humans afflicted with diarrhea, lack microtubules and mitochondria. No mitotic spindle is formed during division of the nucleus, and of course neither mitosis nor meiotic sex is present. Glutathione, a sulfur-containing tripeptide apparently universal in eukaryotes, is also absent in these microbes. The evolutionary appearance of glutathione metabolism may have been correlated with the acquisition of mitochondria (Fahey et al., 1984). *Entamoeba* is probably a genuine relic that appeared before the acquisition of mitochondria and microtubule systems and therefore requires no glutathione metabolism. Glutathione metabolism seems to be required for the regulation of intracellular oxidation-reduction state in meiotic eukaryotes, all of which evolved from ancestral cells containing both mitochondria and microtubules.

Like *Pelomyxa*, *Entamoeba* thrives in low- or no-oxygen environments. Since it is difficult to believe that mitosis would have been entirely lost once it was acquired, both these genera probably evolved before the appearance of mitosis in "main line" eukaryotes. The groups of unified traits known as semes are so important to organisms that evolution has reinforced their presence in the form of integrated genetic systems. It is virtually impossible for a seme to disappear without a trace and, though organisms may revert, they never return completely to their former status. Members of flightless bird species, for example, may tend to lose their wings after long lack of use but they maintain a bone structure and set of wing muscles that clearly reveals a winged ancestry. The flowers may be reduced beyond recognition in flowering plants but they do not disappear completely. On the cellular level the seme of aerobic respiration involves a persistent biochemical and genetic system; in some sulfide-tolerant anaerobic ciliates, for example, rudimentary mitochondria persist. Animals capable of survival in the absence of oxygen still maintain peculiar sorts of respiratory tissue and "anaerobic mitochondria," revealing that they evolved from fully aerobic, oxygen-breathing ancestors. The study of the diminution and loss of semes provides crucial clues to ancestry: mitosis and meiosis are such valuable semes that, once acquired in a lineage, they may have been modified and reduced, but they never disappeared without a trace of their former existence. On the principle that semes never disappear, we must assume that *Pelomyxa* and *Entamoeba* evolved early, before the semes of mitosis. Some protists apparently acquired the spirochete symbionts before acquiring mitochondria (*Pelomyxa* has surface 9 + 2 undulipodia but lacks mitochondria; Whatley and Chapman-Andresen, n.d.), whereas some acquired mitochondria but not spirochetes (*Nanochlorum* lacks microtubules but has mitochondria; Zahn, 1984). These organisms, together with the dinomastigotes, again illustrate the idea that the nuclear membrane, undulipodial motility, and mitochondria originated before chromosomes, mitosis, and meiosis.

# 8 · CHROMOSOME DEPLOYMENT IN MITOSIS

## Microtubules and Their Organizing Centers

### THE MITOTIC APPARATUS

The mitotic apparatus consists at least of the mitotic spindle and often of other microtubular structures such as asters and centrioles as well (fig. 29). Visible in many cells during mitotic division, it is entirely responsible for the movement (called "segregation") of offspring chromatin to the poles of dividing cells, that is, to the incipient offspring cells (fig. 30). In this way mitosis ensures the orderly distribution of at least one set of genes to each offspring cell. Since the absence of genes or their presence in irregular numbers is lethal, the mitotic apparatus is indispensable to cells that divide by mitosis, as nearly all eukaryotic cells do.

Chromatin, fully capable of intense and reversible tight contraction into chromosomes, forms immobile masses (fig. 31). By itself it cannot move toward the poles of cells. Chromatin is not capable of any change in position or translational movement. If it were not for the mitotic spindle, the replicating DNA and protein of chromatin would produce ever-doubling, immobile masses of coiling and uncoiling chromosomes. As we have seen, the mitotic spindle (see fig. 29) is composed of collections of fibers, which are bundles of microtubules, each with the familiar constant diameter of 240 angstroms. These microtubules, composed of tubulin alpha and beta proteins, are homologous to those of the nerve cells of the brain and of all undulipodia.

The mitotic spindle in living cells is such an ephemeral structure that for many years its very existence was doubted. In those cells in which it can be seen it appears during the stages of actual division of the parent cell into offspring cells and disappears after division and during growth. In no cell is it a constant feature. Rather, in all mitotic cells it seems to grow by elongation of fibers. With the aid of the electron microscope this is seen to be the elongation of individual microtubules. The spindle later seems to dissolve, or in some other way to vanish. The electron microscope shows that this is due to the disassembly, or shortening, of these same microtubules. In normal cells the mitotic spindle appears after the chromatin is doubled and condenses into double chromosomes with two chromatids per chromosome. The spindle remains as the chromatids segregate from each other, moving to opposite extremes of the cell. After chromatin segregation the spindle dissolves again, only to reappear after another round of chromatin replication.

In animal cells, especially the large cells of eggs that can be easily observed, the mitotic apparatus can generally be seen to consist of more than the spindle itself. At the extremes of the cell, at the tips of half cones forming the spindle, are structures that with the light microscope appear to be little dots. Because they can be stained with iron hemotoxylin and other stains and can be seen to double at some determined stage in the cell cycle, generally just before or just after cell division, these structures have attracted the attention of biologists as "self-replicating" entities since the 1880s, when they were first described. They have been called "cell centers," "centrioles," "division centers," and other such names, depending on their appearance. A definite relationship between the centrioles of mitosis and the kinetosomes of undulipodia has been recognized since the late nineteenth century, although the details of that relationship were never clear. Now the electron microscope has made it obvious that these are identical structures, except that the kinetosome is always associated with an undulipodium. The centriole is identical in structure to the kinetosome, but has no associated undulipodium. In 1898 two cytologists, the Czech M. von Lenhossek and the Frenchman L. F. Henneguy, argued that the small staining granules in the cell center (centrioles) moved from their position near the nucleus to the cell membrane, where they formed kinetosomes of the cilia. Based on studies of animal epithelium tissue, the Henneguy-Lenhossek theory established that the centriole, located at the mitotic pole, was related to the kinetosome, found at the base of all undulipodia, long before it was known that both these structures are composed of microtubules in the 9 + 0 array. Although the details of replication and development of kinetosomes from centrioles are complicated, there is no doubt that Henneguy and Lenhossek were fundamentally correct: the 9 + 0 centriole and the 9 + 0 kinetosome, products of replication in the cell that appear independently of the nucleus, mitochondria, or plastids, are interchangeable. The centriole at the poles in mitosis can actually grow a shaft (the axoneme)

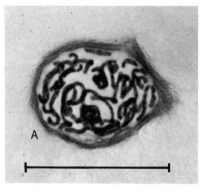

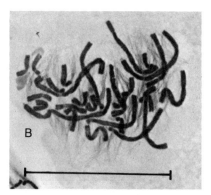

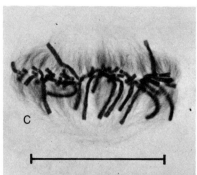

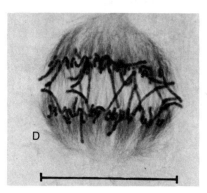

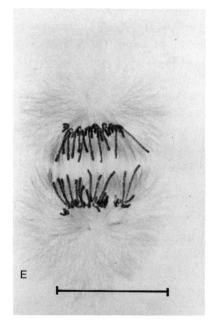

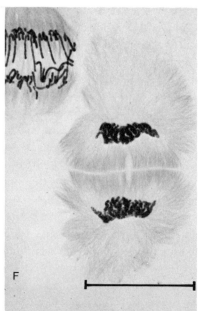

126

and become, thereby, a kinetosome. Centrioles and kinetosomes are the same structures with the same evolutionary origin (Wheatley, 1982).

Both kinetosomes and centrioles are a quarter of a micrometer (250 nm) wide. But the two structures vary in length as they grow outward in their ninefold symmetry. Both may grow from the same length as their width to up to 3 or 4 micrometers long. The 9 + 0 arrangement of microtubules is visible only with the electron microscope. These structures are absent at the poles of division in nearly all plant cells and are extremely difficult to see in certain animal cells; in fungi they are entirely absent. Thus the function, if any, of the 9 + 0 centriole itself (sometimes called the "centriole pinwheel") has always been a subject of debate. There are many lines of evidence suggesting that centrioles are not universally or directly necessary for mitotic cell division (Pickett-Heaps, 1974; Wheatley, 1982).

Some parabasalids (fig. 32), the hypermastigotes, provide us with a major part of our story. These "hairy men," as they are familiarly referred to, have certain peculiar structures that regularly appear at the extremes of the dividing cell. They are enormous and thus unambiguously observable. These long, unusual, hypermastigote structures are more rod-shaped than dot-shaped. But because they appear during cell division at the edges of a fibrous mitotic spindle, L. R. Cleveland, in a series of studies over several decades, called them centrioles and treated them as equivalents to the tiny, staining dots in animal cells. Although Cleveland's interpretation has been severely criticized (Fulton, 1971), because the hypermastigote rods are not equal to the pinwheel centrioles at the poles of animal cells, Cleveland was correct in considering them functional centrioles in the sense of microtubule organizing centers involved in cell division.

**Fig. 29.** Stages of mitosis. Light micrograph of cells from the endosperm (tissue supporting the embryo in the development of the seed) of *Haemanthus katherinae* Bak. (S. African blood lily). Bar = 30 micrometers. A. *Prophase*. A dense accumulation of microtubules (MTs) surrounds the nucleus. All MTs in the cytoplasm are disassembled. B. *Prometaphase*. Kinetochores migrate to the equator. Individual chromosomes with their kinetochore fibers are well seen. C. *Metaphase*. Kinetochores are arranged in one plane, forming the metaphase plate. D. and E. *Anaphase*. Chromosomes migrate to the poles. Some individual kinetochore fibers are visible, and also some MTs in the interzone. Newly formed polar MTs form asterlike structures around the poles. Polar MTs also between the trailing chromosome arms toward the equator. F. *Telophase* (with a second cell in anaphase). Chromosomes condense at the poles. Phragmoplast is formed by fusion of sets of new MTs growing from opposite poles. (Courtesy of A. Bajer and J. Mole-Bajer, University of Oregon.)

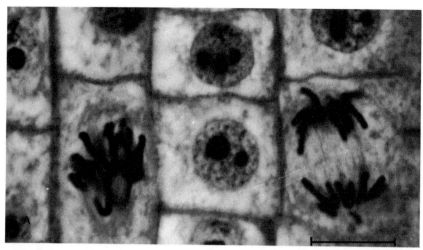

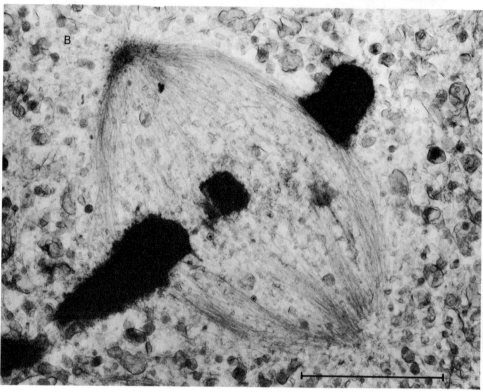

**Fig. 30.** Mitotic spindle and chromosomes in mitosis. A. Light micrograph (bar = 10 micrometers). B. Electron micrograph (bar = 5 micrometers). (Courtesy of Richard McIntosh, University of Colorado.)

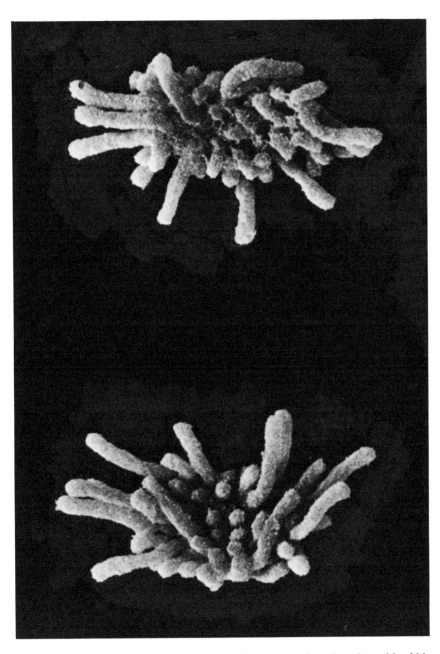

**Fig. 31.** Chromosomes in anaphase. Micrograph of *Haemanthus*, the African blood lily, in mitosis. No mitotic spindle can be seen but its microtubules were responsible for movement of the chromosomes to the poles of the cell. The distance from pole to pole is about 30 micrometers. See fig. 29. (Courtesy of M. Laane, University of Oslo.)

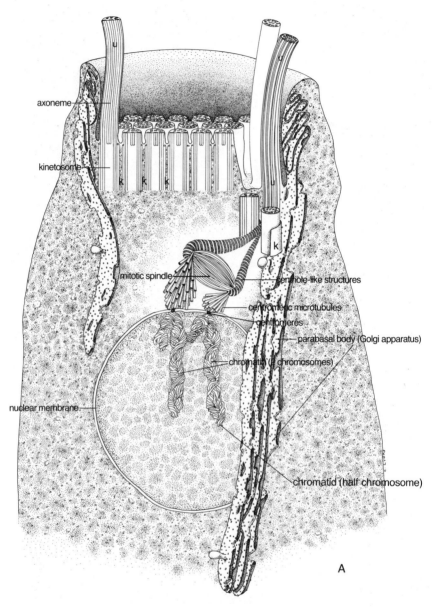

**Fig. 32.** Parabasalids. A. Structure of the anterior portion of the parabasalid *Lophomonas striata* showing the relationship between undulipodia (u) and mitotic apparatus. (Drawing by Christie Lyons after Hollande and Caruette Valentin, 1972.)

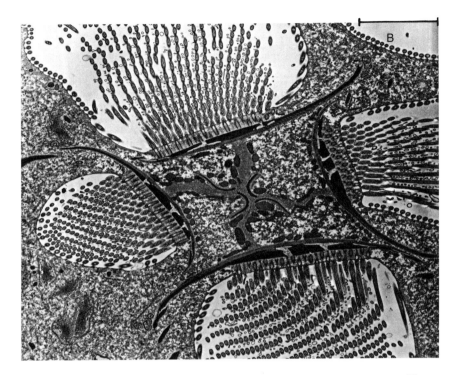

**Fig. 32.** Parabasalids. B. Electron micrograph showing transverse sections of hundreds of undulipodia in the parabasalid *Staurojoenina* sp. (bar = 5 micrometers). (Courtesy of Dr. David Chase.)

In the early 1960s, when glutaraldehyde became generally available as an effective fixative of microtubules, ultrastructural studies were begun. They revealed the true nature of Cleveland's "centrioles," and of many other "division centers" of protists that in live cells were seen only as dots or rods. Such "division centers" vary enormously in detail, but they are united by their conspicuous presence during cell division. They serve as places of initiation of elongation of microtubules. In particular, the so-called pinwheel centrioles and other more complex division centers are associated with even tinier, amorphous structures. These pericentriolar dense bodies serve as organizers of mitotic spindles and other microtubules. The pericentriolar bodies are seen in electron micrographs as dense bodies associated with 9 + 0 centrioles. It seems that the microtubule organizing center—not the centriole itself—is capable of replication. A product of this replication can then develop into one or another of a series of microtubular structures, visible or not at various levels of microscopy. The centriole pinwheel and centriolar rods are not themselves microtubule organizing centers but are their products.

## MICROTUBULE ORGANIZING CENTERS

MTOCs, first recognized as points in eukaryotic cells at which microtubular structures develop by outgrowth at highly determined times and places, are responsible for the asymmetry of eukaryotic cells. Unresolvable as clearly defined structures, they must be inferred by the appearance and growth of clearly defined microtubule-based cell organelles in living cells.

As work with MTOCs expands through the correlation between biochemistry and morphology (for example, that between images at the electron microscopic level and the appearance of protein-specific antibodies that can be identified, separated, and stained), the concept of MTOC becomes more clear. In at least one case, that of the pericentriolar material, MTOCs are RNP—that is, ribonucleoprotein in mammalian cells (Calarco-Gillam et al., 1983). The appearance of spindle fibers, which in some cells grow out from specific points (again seen as dots in a light microscopic preparation) on the surface of chromosomes, implies that these dots are MTOCs. The "dots" at which the spindle microtubules connect to the chromosomes are indeed MTOCs, although they have many aliases in the biological literature. Called "spindle fiber attachments," "kinetochores," or "centromeres," these dots too, like centrioles, have been claimed since the late nineteenth century to be self-replicating.

Chromatids, which as we have seen are half chromosomes, become chromosomes as they segregate from each other in the mitotic process. Chromatids always move through the cell kinetochore first. The kinetochore is attached to the microtubules of the mitotic spindle, but if this attachment is interrupted or broken, the chromosome is immobilized. That is, chromosome movement is passive. Chromosomes are pushed (or pulled or carried) through the cell by the mitotic spindle apparatus. The kinetochores, located on the chromosomes, are places from which microtubules emanate and grow.

Microtubule organizing centers are not limited to the poles (where they are associated with centrioles) and surfaces of the chromosomes (where they are associated with kinetochores). Such centers are also at the bases of axonemes of axopods (the "feet" of heliozoans; fig. 33). MTOCs can be seen faintly at the poles of dividing yeast nuclei (fig. 34). Enormous complex "division centers" first described by Belar (1926) are clearly magnificently structured microtubule organizing centers for cell division in *Dimorpha* (and *Tetradimorpha*; Brugerolle and Mignot, 1984). MTOCs are assumed to be at the bases of suctorian tentacles and the axostylar rods of pyrsonymphids, and

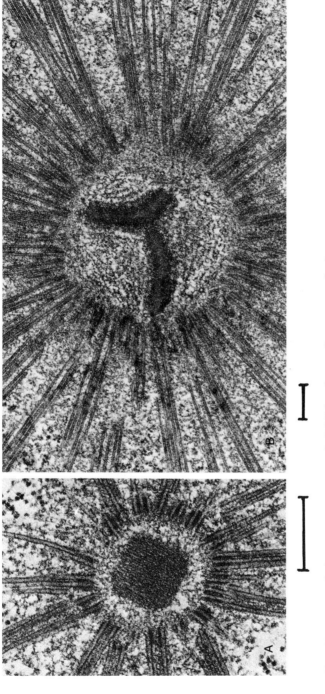

**Fig. 33.** Microtubule organizing centers (MTOCs) at bases of axonemes of a heliozoan. A. Centroplast of *Heterophrys marina* (bar = 0.5 micrometers). B. Centroplast of *Heterophrys magna* (bar = 0.5 micrometers). (From Bardele, 1977b; courtesy of Christian F. Bardele, University of Tübingen.)

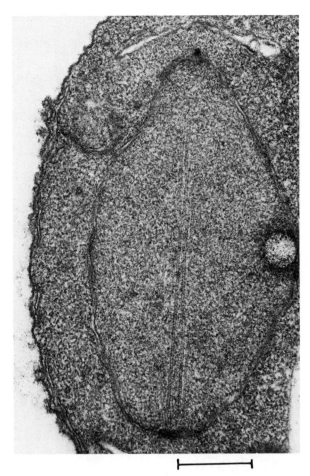

**Fig. 34.** MTOCs in dividing nucleus of a yeast cell. Electron micrograph of meiosis II phase in *Schizosaccharomyces octosporus (bar = 1 micrometer )*. (Courtesy of Mary Lou Ashton, York University.)

they are implicated wherever in eukaryotic cells extensive arrays of patterned microtubules are found (as in *Pyrsonympha, Actinophrys, Acanthocystis*, and *Allogromia;* fig. 35). They are associated with protruding and intracellular structures of many kinds. That these MTOCs play a crucial role in the development of cell asymmetries has been easily demonstrated in a large number of cells (table 10).

The search for the self-replicating entity at the very center of these microtubule organizing centers themselves has led to a variety of generalizations. The first is that the minimal microtubule organizing center is seen as fuzz; it is a dense, granular or granular-fibrous material obvious only after the microtubules begin to emanate from it. Invisible with the light microscope, it can in many cases only be inferred even with an electron microscope. Nevertheless, looking through the electron microscope it becomes apparent that the MTOCs are embedded in the nuclear membrane in many protists.

A second generalization is that microtubule organizing centers, however minimal their size and morphology, are present wherever and whenever microtubules begin to grow in eukaryotic cells.

Third, microtubule organizing centers are responsible for the production of a number of almost ubiquitously distributed cell organelles, such as centrioles, kinetosomes (centrioles with associated axonemes, such as the kinetosomes of the ciliate cortex), the asters of eggs, and kinetochores (spindle attachment sites). The microtubule centers should be conceived of as the genetic basis for the presence of these structures.

Finally, although microtubule organizing centers apparently are capable of self-replication and the elaboration of many sorts of products, such as those mentioned above, there is no evidence that they are composed of DNA. Several experiments indicate, however, that microtubule organizing centers are composed of RNA rigidly held in place and complexed with protein (Hartman, Puma, and Gurney, 1974; Younger et al., 1972; Dippell, 1976).

## MICROTUBULE CENTERS AND MITOSIS

In the standard mitosis of animals, plants, fungi, and some protoctists, an interaction between the microtubule centers and chromatin achieves an equal segregation of genetic material to offspring cells. An enormous diversity in the details of mitosis has been reported (Heath, 1980a, 1980b), yet all involve interaction of microtubules with chromatin. In some the nuclear membrane is directly involved in the process of chromatin segregation, in others the nuclear membrane dissolves and plays no role. The number of microtubules in the mitotic spindle and the extent of their development vary enormously. The minimal number (found in some yeasts) involves a single microtubule inserted into a strand of chromatin, one for each of the six chromosomes.

The common denominators of all standard mitotic events include the replication of chromatin prior to the onset of mitosis, the formation of chromosomes by the "condensation" of DNA and protein, and the presence of an MTOC, which determines at least one kinetochore for each of the chromosomes. The kinetochore is generally on the chromosome, but it may, as in some hypermastigotes, be attached to the spindle and only later, in the course of the division itself, establish contact with the chromosomes.

The replication of the MTOC leading to the duplication of each

Drawings
of protist
structure

*Acanthocystis chaetophora*

mt

Electron
micrographs
showing
microtubules

*Heterophrys elati*

Drawings (left)
and electron
micrographs
(right) showing
microtubule arrays

*Raphidiophrys ambigua*

**Fig. 35.**  Uses of microtubular arrays in protists; mt = structure underlain by microtubules. In heliozoan axopods (*Acanthocystis, Raphidiophrys, Echinosphaerium*), microtubules are feeding and prey-capture organelles. In foraminiferan rhizopods, such
as those of *Allogromia,* they are active in food capture, in locomotion, as intracellular
conduits for mitochondria and other particles, and in calcium carbonate shell deposi-
tion. In pyrsonymphids microtubules comprise the axostyles such as this one in

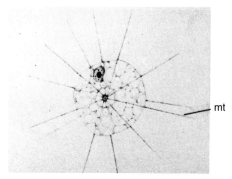

*Actinophrys*

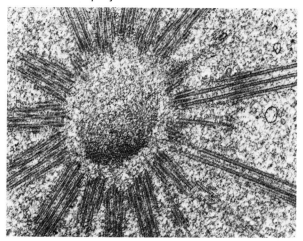

*Acanthocystis penardi*

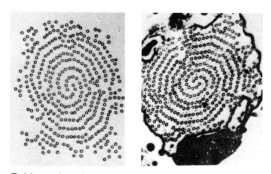

*Echinosphaerium*

*Pyrsonympha;* they undergo vigorous nontranslational movements and may function as "stirrers." In other pyrsonymphids (*Saccinobaculus, Oxymonas, Notila*) the axostyles fuse during the sexual cycle and serve to unite the male and female gamete pronuclei. Magnification can be estimated from the constancy of the microtubule diameters at 24 nanometers.

Fig. 35 *(continued)*

FORAMINIFERAN

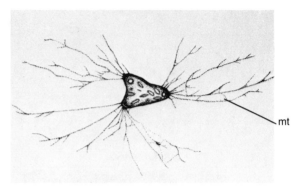

Drawings
of protist
structure

*Allogromia*

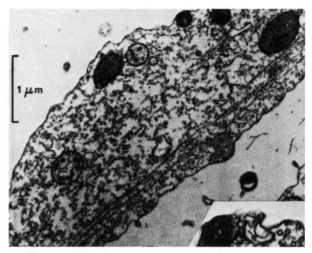

Electron
micrographs
showing
microtubules

*Allogromia*

Drawings (left)
and electron
micrographs
(right) showing
microtubule arrays

*Allogromia*

# PYRSONYMPHID

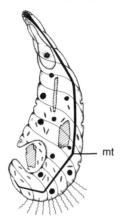

*Pyrsonympha*

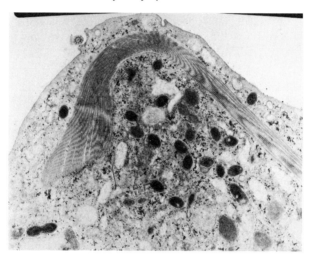

*Pyrsonympha*

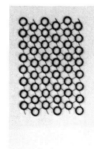 

*Pyrsonympha*

139

**Table 10.**  Examples of Microtubule-Based Morphogenetic Processes in Protoctists, Animals, and Plants

| BIOLOGICAL SYSTEM | DESCRIPTION OF PROCESSES |
| --- | --- |
| PROTOCTISTS | |
| *Carchesium,* colonial, peritrichous ciliate | Stalk development |
| *Stentor,* heterotrichous ciliate | Oral membranellar band morphogenesis (many undulipodia) |
| *Tetrahymena,* a holotrichous ciliate | Macronuclear division and cell growth |
| *Tokophrya,* suctorian ciliate | Micronuclear and macronuclear division |
| *(Echinosphaerium) Actinosphaerium,* heliozoan | Development of retractable axonemes |
| *Nitella* (charophyte alga) | Sperm-tail morphogenesis, branched undulipodia induced by microtubule inhibitors, ordered microtubules required for normal morphogenesis |
| ANIMALS | |
| Cockroach, campaniform tactile spines (mechanoreceptors) | Microtubule loss correlated with loss of response to stimuli after 1–2 hours |
| *Rhodnius* (squid) developing eye | Outgrowth of lens primordium requiring microtubules |
| Mouse macrophage in culture | Pseudopodlike processes and correlated amoeboid motion in gliding cells |
| Spinal ganglion | Axonal reelongation |
| Chinese hamster cells in culture | Formation of continuous spindle fibers |
| PLANTS | |
| *Allium,* onion root meristem | Spindle formation, many chemical inhibitors of microtubules tested |
| *Triticum,* wheat | Somatic association of homologous chromosomes requiring microtubules |

*Note:* See Dustin, 1978, and Margulis, 1974, for details.

140

kinetochore, the establishment of contact between the kinetochores of each chromosome and the mitotic spindle microtubules, and the segregation of chromatids from each other are also regular mitotic occurrences. In the segregation of chromatids, half the chromosomes move during the formation of two offspring cells toward each extreme of the parent cell. The actual movement of the chromatids requires the presence of dynamic microtubules. In some cases it can be shown that the segregation of chromatin results directly from the elongation of the microtubules comprising the spindle (Ris, 1975; Heath, 1980a, 1980b). In other cases the controlled disassembly of microtubules seems to be involved (Goode, 1981).

One final mitotic requirement is cytokinesis: the division of the cytoplasm that results in two masses of chromosomes in two offspring cells each of which are genetically identical to the parent cell. Hence, as was clearly recognized by E. B. Wilson (1925) and several of the authors whose work he summarized, the mitotic apparatus processes of mitosis are entirely separable from the chromosomal DNA aspects. Wilson stated:

> [A] fundamental dualism exists in the phenomena of mitosis, the origin and transformation of the achromatic figure [mitotic spindle] being in large measure independent of those occurring in the chromatic elements [chromatin]. Mitosis consists, in fact, of two closely correlated but separable series of events. This conclusion greatly facilitates an experimental analysis of the general problem. (p. 178)

## DOUBLE ORIGIN OF MITOSIS

The least sturdy hypothesis of the serial endosymbiotic theory of the origin of eukaryotic cells is that of the origin of undulipodia from free-living bacteria, most likely from spirochetes. But this hypothesis is strengthened by its considerable explanatory force in reconstructing the origin of mitosis and meiotic sex. The hypothesis that undulipodia were originally motile symbionts that became organelles of motility greatly simplifies our analysis of the origin of sex. We trace here the steps in the origin of meiotic sex based on the spirochete hypothesis.

The undulating movements of the spirochetes conferred selective advantage on the early eukaryotic complex. The associated spirochetes developed permanent attachments to their hosts. With time they became entirely dependent on the metabolic products of the hosts. With each generation the spirochetes reproduced in a ratio approximately constant with that of their hosts and thus the complex maintained a roughly constant number of spirochetes per host (later equivalent to organelles per cell). If the spirochetes had increased disproportionally, the undulipodial elements would have overrun and destroyed the host cells that they had invaded. The continued presence of motile, replicating spirochetes, in the process of becoming obligate symbionts and then organelles in the early eukaryotic complexes, was a preadaptation for the emergence of mitosis. The intracellular presence of motile bacteria in the first eukaryotes provided an internal transportation system of unprecedented complexity and potential. The disassembly and redeployment of the former spirochete components, the continued replication of the nucleic acids responsible for these components, led to the emergence of an entity of far greater dynamic organization than any bacterial cell. The complex motility systems of eukaryotes (pinocytosis, phagocytosis, exocytosis, endocytosis, and mitosis) hypothetically derived from this original merger. The elaborate processing of nucleic acids, the large ribosomes, and the incessant internal activity of eukaryotes behave as products of the composite ancestry.

We see today in the eukaryotic cell the results of this ancient microbial community interaction between host and motile symbionts. We can only surmise the events that led from a symbiotic complex that included motile spirochetes to a mitotic cell. The spirochetes merged entirely with their hosts; their facility for replication and movement was taken over by the "central government" of the hosts. Spirochetes and their attachment sites became kinetids, spirochetes transformed to undulipodia. The intracellular motility of eukaryotes, their kinetodesmal fibers (striated rootlets), microfilaments, and microtubules are acquired morphologies with accompanying physiologies. With the intracellular acquisition of the undulipodia, once spirochetes, came flourishing possibilities for intracellular movement. These possibilities included mitosis and eventually complex, multicellular development. Hungry, lithe, and speedy spirochetes merged with huge, well-nourished, but torpid hosts. Many recombinant structures appeared: kinetochores, centrioles, kinetids with all their associated fibers, and double membranes are all components that should be traceable to dual origins. The obligation to autopoiesy and replication of course persisted, but spirochete motility systems were coopted for, among other things, the equal delivery of host genomes to the extremes of the parent cells, which, after cytokinesis, became offspring cells.

The key concept here is our hypothesis that microtubule organizing centers of eukaryotes are the remnants of the spirochete genomes. In the course of evolution the obligate replication, associated protein synthesis, and assembly of the MTOCs required for the persistence of the motility complex was utilized in the origin of many new structures, among them kinetochores, centrioles, and kinetosomes. The MTOCs—spirochetes stripped of nearly all but their motive and replicative powers—are alien in origin but are now wholly integral parts of all eukaryotes. The duality of mitosis is attributable to its double ancestry: the microtubule-motile complex (from the spirochetes) and the nucleochromatin complex (originally from the host and other organellar genomes). From the two autopoietic systems and their continued, correlated rates of replication come the central features of mitosis. Mitosis and meiotic sex arose as organizing phenomena from the mire and confusion of microbial community life.

The different heritages can be traced. The duplication of chromatin before the onset of mitosis originally came from the host. The formation of chromosomes by the condensation of chromatin evolved as the host co-opted and utilized the spirochete motile protein system. The presence and replication of an MTOC is owed to the remnant spirochete genome, which now determines a duplicating kinetochore for the division of each chromosome.

The establishment of contact between the two kinds of MTOCs—that is, between the kinetochores of each chromosome and the centrioles or equivalent structures that form the mitotic spindle microtubules—is conserved from the Archean processes that established spirochete-host attachment sites. The segregation of the chromosomes into the opposite extremes of the parent cell in the formation of two offspring cells represents the use by the host of motility systems originally belonging exclusively to spirochetes. The ancient spirochete motility apparatus is used in eukaryotes to segregate chromatin instead of DNA on membrane, as occurs in bacteria and some protists.

Cytokinesis, the division of the cytoplasm that results in two masses of chromosomes in two offspring cells genetically identical to the parent cell, is another example of the utilization by the host of the original spirochete motility systems.

The hypothesis of the equivalence of MTOCs and remnant spirochete genomes provides an integrated explanation of the origin of mitosis and, as we shall see, of meiosis. It leads directly to certain testable conclusions. One of these is that certain free-living spirochetes will contain proteins homologous to the motile proteins of eukaryotes (tubulins, dynein, and other undulipodial proteins). Such proteins will not be present in the codescendant microbes thought to be ancestors to the host (*Thermoplasma*), to mitochondria (*Paracoccus, Bdellovibrio, Rhodopseudomonas*; Sagan et al., 1985), or to plastids (*Prochloron, Syneococcus, Heliobacterium*; Margulis and Obar, 1985).

If MTOCs are indeed the evolved offspring of ancient motile bacteria, then certain free-living spirochetes will contain DNA homologous to the RNA of MTOCs, for example, to the RNA found associated with kinetosomal and kinetochoric replication. Also on the macromolecular front, homologies will be found between enzymes (for example, nucleic acid polymerases) in certain free-living spirochetes and those of mitotic eukaryotes. These polymerases will be less homologous to those of the codescendant microbes thought to be ancestors of the nucleocytoplasm, mitochondria, or plastids.

Two other predictions follow. Unique homologies, not represented in codescendants of the nucleocytoplasm (*Thermoplasma*), will be discovered between the RNA and protein of the ribosomes of certain spirochetes and those of mitotic eukaryotes. In addition, ribosomes will not be uniform in eukaryotes because different lineages will have retained different proteins and RNAs of the original spirochetes.

## PROTISTS AND FIRST MITOSIS

If these conclusions are borne out by evidence, then the first mitotic protists must have been complexes of host-mitochondria infected with motile spirochetes that had become undulipodia. The beating undulipodia internal to the cell membrane became swimming organelles used to propel the complex from place to place, conferring a motility advantage not entirely unlike the advantageous role that horses played in the spread of the Indo-European peoples. In many lineages of protists the undulipodia, having entered the cell for feeding purposes, were retained inside it and exploited to distribute the replicated DNA of both host and former spirochete. The products of chromatin replication in the host attached to the segregating undulipodia. Both sets of genomes together moved to the extremes of the complex. The complex, since it acquired mitochondria and attached spirochetes, had been a microbial community. It was now in the process of becoming a cell (see fig. 21).

In such cells and their multicelled descendants, including the mammal *Homo sapiens*, undulipodial movement was not possible during cell division. Even today all animal cells with undulipodia, such as the ciliated epithelia of human oviducts and tracheae, must divide *before* they produce undulipodia. In the protist lineages that were to evolve into protoctists, and later into animals, selection pressures led to larger motile cells capable of evenly distributing their tangles of nucleic acid.

In these lineages chromosomes associated with spirochete remnant genomes. Synchrony was selected for. Each time chromatin replicated, so did the remnant genome of a spirochete, providing each chromosome with a kinetochore. Errors were made: spirochetal remnants replicated more quickly and less quickly than their chromosomal counterparts. In certain lineages when kinetochores replicated faster than their chromosomal counterparts, chromosomes fissioned to become half chromosomes. Errors of more rapid replication led to karyotypic fissioning: a spontaneous dividing of large chromosomes into small chromosomes that has been considered important in the evolution of many species of mammals (Todd, 1970; Giusto and Margulis, 1981). In karyotypic fissioning and similar events some spirochetal remnants hesitated, their associated chromosomes failed to enter offspring cells, and the offspring cells thus died. Relentless selection refined mitosis. The ritualization of mitosis occurred in many lineages: in chlorophyte ancestors to the plants, conjugating protoctist ancestors to the fungi and red algae, and the ancestors of the diatoms and ciliates. Such distribution patterns of spirochete remnants and genetic material occurred in dinomastigotes before the emergence of true chromatin in that lineage. But ritualization of mitosis was far from universal: for example, it never occurred in the ancestors to amoebae, kinetoplastids (bodos and trypanosomes), or euglenids.

Refined mitosis is a prerequisite to meiosis. Mitosis was never universal in protoctists, nor is it universal now. And only in a few mitotic protoctist lineages, most likely those which had evolved by the late Proterozoic Eon, did meiotic sexuality ever evolve. As we shall see, these lineages are our heritage.

# 9 • CANNIBALISM AND OTHER MERGERS

## Protistan Dilemma of Doubleness

### NO PLOIDY, ANEUPLOIDY, AND HAPLOIDY

Meiotic sex encompasses the halving and doubling of the number of chromosomes. The doubling of chromosomes, into what is known as the diploid state of two sets of chromosomes, begins with fertilization. In this chapter we explore the possibility that protists originally cannibalized other protists but did not digest them. This would have doubled the number of chromosomes without the fertilization that later became an established part of the life cycle of some eukaryotes.

In the last century August Weismann developed the concept of "ploidy," referring to the number of sets of chromosomes in a plant or animal cell. (The word derives from the Greek *ploos*, meaning "fold," as in *twofold*. Cells with a single set are haploid, those with two sets are diploid, with four sets tetraploid, and so on.) Animals are considered diploid because their body cells (that is, all cells except sperm and eggs) contain two sets of chromosomes. Their "sex cells" or gametes, however, are haploid. Gametes are necessarily haploid because they fuse during fertilization. If each were diploid, the new cell (zygote) would have four sets of chromosomes rather than two. Indeed, this does occasionally happen, resulting in a fourfold or tetraploid zygote.

Any single complete set of chromosomes is referred to as euploidy ("true" ploidy); an uneven number—too many or two few—produces aneuploidy (that is, *not* euploidy), often a deadly state.

The fusion of two haploid gametes during fertilization naturally produces a diploid zygote. The organism that grows from this zygote must, however, again produce haploid gametes. The reduction from diploidy to haploidy is accompanied by meiosis. This is a series of one or two cell divisions that produce haploid offspring cells from diploid parent cells. Meiosis is unnecessary in the production of gametes by already haploid protoctists and plants. Protoctists such as apicomplexans and plants such as mosses produce haploid gametes by mitosis; these gametes fuse to form diploid zygotes that grow into the diploid stage of the organisms. Many plants produce haploid spores, which are resistant to desiccation. These then germinate into haploid plants. (The haploid plants that grow from spores of mosses or ferns are called "gametophytes.") In summary, the body cells of plants may be either haploid or diploid. The only rule is that fertilization must be followed by meiosis before fertilization can occur again.

The field of biology that investigates the relation between organisms and their chromosomes is cytogenetics. One of its major concerns is karyology, the study of karyotypes. The karyotype of an organism is the number and morphology of its chromosomes. Cytogenetics deals only with eukaryotes. Neither karyotype nor ploidy is a concept applicable to prokaryotes, since prokaryotic DNA is not organized into chromatin and since prokaryotes lack chromosomes. Unfortunately, among microbiologists the term *haploid* has been applied loosely to the bacterial genome, referring to the fact that bacteria generally have a single set of genes. In fact, the prokaryotic synthesis of DNA may go through several rounds, with the result that growing bacteria, unable to synthesize cell walls rapidly enough, sometimes contain several sets of genes. But to retain its usefulness, the concept of ploidy, which indicates the chromosomal organization of nucleated cells, should be restricted to mitotic organisms.

The majority of animals and plants, then, because of the nature of their life cycles, alternate between diploidy and haploidy. The ploidy of fungi has been more difficult to determine, because many of these organisms have no visible chromosomes. Others have tiny and often numerous chromosomes. Fungi, in the strict sense limited to zygomycotes, ascomycotes, and basidiomycotes, all grow from desiccation-resistant propagules (spores), cells with haploid nuclei and hardened walls composed of chitin. Many fungi show a sexual stage in their life cycles resulting from fertilization. The cytoplasm fuses, but not the nuclei. Thus fungus cells do not become diploid but become dikaryotic (literally "two-seeded"); their cells contain two different nuclei surrounded by a common cytoplasm.

A single spore can germinate to form long, branched threads (hyphae) made of strands of connected haploid cells. In many fungi, such as the common mushroom (*Agaricus*), hyphae of differing mating types meet and fuse. However, haploid nuclei of different parental origins remain in the common cytoplasm without fusing. In both ascomycotes and basidiomycotes no karyogamy may occur for days to more than years. The simultaneous presence of two different kinds of haploid nuclei in the common cytoplasm is *dikaryosis*. Genetically this is the product of a sexual fusion. It is comparable to plant and animal diploidy in the sense that two parental nuclei, each comprising one half the genetic content of the cell, are present at one time, making up the new genome. Cytologically, however, since the nuclei do not fuse to form a diploid zygotic nucleus, fungi are haploid organisms. In summary, bacteria show no ploidy, fungi are haploid, animals are diploid, and plants alternate haploidy and diploidy. What, then, is the ploidy of protoctists, the ancestors of animals, plants, and fungi? How did ploidy, the regularization of chromosome sets, arise in the history of life?

## PROTOCTISTS AND PLOIDY

Some protoctists, such as the single-celled chlorophyte *Chlamydomonas* and its extant relatives, presumed to be multicellular descendants of *Chlamydomonas*-type cells (for example, *Eudorina*, *Pandorina*, and *Volvox*), are all-haploid organisms that fundamentally display the same life cycle as fungi. From hard-walled cysts emerge haploid, motile, green single cells that reproduce by mitotic cell division. If these organisms fuse in fertilization they immediately undergo meiosis to restore the haploid number of chromosomes in each cell. *Chlamydomonas*, like many of its green algal relatives and other protoctists, apparently must grow in the haploid state.

In some laboratory cultures of protoctists the number of chromosomes and quantity of DNA per cell of the same species (for example, euglenids or dinomastigotes) vary enormously. The addition of phosphorus to the culture medium, for example, may drastically alter the amount of DNA, despite the fact that it is of a given species at a given time. There is no obvious and fixed ploidy. Like bacteria, some protoctists, for example, many amoebae and trypanosomes, have no set number of chromosomes and therefore no ploidy. Their chromosomes show no regular morphology and organization that is definable as a clear karyotype. In still other protoctists, such as diatoms, chromosomes as obvious and clear as those in any animal can be seen. *Lithodesmium* and *Surirella*, for example, are diploid; meiosis occurs prior to the formation of sperm and egg just as its does in animals (E. B. Wilson, 1925; Pickett-Heaps, Tippett, and Andreozzi, 1978; Lauterborn, 1896).

The variations in ploidy found in protoctists—from none at all to the typical sorts of karyotypes observed in animals and plants—are best explained by the concept that ploidy evolved in these organisms. Euploidy is a consequence of orderly mitotic division. We have seen in chapter 8 how mitosis may have evolved. It is not surprising that many protoctists are not rigorously mitotic: as long as there are no continual restraints on the constancy of chromosome number, deviations from the mitotic and haploid rule may be tolerated. Euploidy, however, has an obligatory correlate in a meiotic-fertilization life cycle. Once meiosis-fertilization evolved, mitosis itself could tolerate very few deviations from the rule that chromosome numbers be preserved.

Since virtually all animals, plants, and fungi are euploids with cells that undergo regular mitosis, the mitotic process itself must have evolved in organisms ancestral to these three lineages. It seems obvious from the extent of the variations on the mitotic theme that mitosis (in haploid organisms), and eventually meiosis (haploidy alternating with diploidy), evolved in protoctists (Heath, 1980b). How? Why? Asking about the origin and evolution of regulated euploidy is another way to address the question of the origin of meiotic sex. The first step in this development can easily be imagined.

## CANNIBALISM

What are the threats to a population of protoctists, whose members are loosely mitotic (in the sense that their chromosomes are distributed in sets to offspring cells by MTOCs)? They are, of course, the eternal threats to heterotrophic organisms: starvation and desiccation. Autopoiesy, as we know, requires a constant supply of organic compounds and water.

Just as populations of humans subject to prolonged intervals of starvation have been known to resort to cannibalism—that is, to eating members of their own species—so have many protoctists. *Blepharisma*, a huge ciliate whose single cell may be a millimeter long, has been known to ingest other *Blepharisma* when threatened (Giese, 1973). Cannibalism has been seen in *Stentor* (another large ciliate) and in various amoebae (Page, 1983). L. R. Cleveland reported this sort of behavior in hypermastigotes. The outcome of these cannibalistic encounters has rarely been described. In some cases the ingestion succeeds, and one organism digests the other. In some cases the ingested partner is rejected and expelled into the medium. In at least one case, reported by Cleveland (1947), cannibalism led to fusion and the formation of a double cell: two nuclei, two sets of chromosomes, and two sets of MTOCs (called "extranuclear organelles" at the time). As long as the doubled organism is selected for and single organisms die, great pressures will lead to equal rates of reproduction of the components of doubled cells.

Although protists lack an immune system, they are capable of protecting the contents of their own cells from autodigestion. They also tend to have a rich assortment of digestive enzymes. It is easy to imagine many possible outcomes of a cannibalistic encounter between early protists. The devoured protist, resistant to its own enzymes, might also be resistant to the enzymes of its predator. It might then have survived for any number of reasons. No doubt in some cases there would have been genetic differences in a cannibalized pair. If these genetic differences, no matter how small, conferred on the grisly couple an advantage over their uncoupled conspecifics, the doubled form would persist. Such a doubled form might outcompete its single neighbors because it would tend to have less surface area per volume and perhaps might have been that much more able to tolerate desiccation or starvation. The result would have been a fused protist cell, incipiently diploid, the hardy product of indigestion (ingestion followed by subsequent failure of digestion). If, however, conditions had been optimum for the haploid, such a duplex cell, most likely derived from a healthy haploid, would be better off in its original haploid state. Should the harsh conditions that spurred diploidy change, selection pressures would tend to return the diploid to its former haploid state.

## OTHER MERGERS

Watery protists, perturbed by constant hunger and thirst, could have doubled their single set of chromosomes without cannibalism. Cytologists are always impressed with the frequency and casualness of error in cell biological processes. Sperm have been observed that have two or three tails rather than one. Supernumerary undulipodia—up to sixteen—can be produced simply by temperature shock in protists such as *Naeglaria*, an amoebomastigote (Dingle, 1979). Mates that fuse by threes instead of twos are known; members of the species *Leptospironympha*, for example, mate by threes in one of Cleveland's fine films (ca. 1950s) on the symbionts of termites and wood-eating cockroaches. Zygotes of frog eggs are formed, the zygote's cells divide, and the cytoplasm that should have been included in the developing embryo remains unincorporated.

Such living mistakes are common. To err is more than human, it is biological. One common error today is the failure of divided nuclei to separate into two offspring cells. In early communities of protists, as now, nuclei were no doubt produced with varying numbers of chromosomes, often with the optimal haploid number: one copy of each gene in each cell. Inhibitors to cytokinesis (cytoplasmic cell division) must have abounded: a sudden chill, an influx of cold water, the appearance of a toxin in the water. The protist arrested in cytokinesis after karyokinesis (nuclear cell division) was converted into a dikaryon: such a cell contained (like so many fungi and protists of today) two nuclei in a common cytoplasm. If the nuclei of such a cell were to fuse, then this cell too would become a functional diploid with the imperative to be relieved of its burden of diploidy under certain conditions.

No doubt the tendency to fuse did not abruptly begin or cease. Many cells probably became fat by cannibalism or failure to divide by cytokinesis. At one point, probably in the late Proterozoic Eon after the atmosphere had become well oxygenated, but before the appearance of the Ediacaran animals (about 750 million years ago), the waters were replete with large and loosely ploid protists of many kinds. Doubled and tripled forms were probably common. Having abandoned the haploid state for at least doubleness in order to survive, these protists were well on their way to becoming our microbial ancestors. But doubling had not solved their dilemma. Many were faced with pressures to return to an undoubled state, because the streamlined haploid cell, with only a single copy of each gene per cell, in some cases must have been capable of leaving more offspring. How the relief of diploidy was achieved and regularized is another question first worked out by Cleveland (1947). The story of the relief from doubleness is a critical step in the origin of meiotic sex.

## FROM CANNIBALISM TO FERTILIZATION

Cannibalism or the failure of cytokinesis may lead to doubleness, but such processes alone cannot account for fertilization. In fertilization one haploid cell must recognize and fuse with another. Such mate recognition usually involves a combination of chemical and behavioral clues. The result of mating activities leads to a fusion of the cells to form the dikaryon (common cytoplasm containing two differently derived nuclei). This process, called "syngamy," is followed by the fusion of the nuclei known as "pronuclei" prior to fusion. The process, known as "karyogamy," forms the new diploid nucleus.

Programmed fertilization followed by meiosis did not occur by one or a few mutations. Meiosis itself is a multistep process. So is fertilization. Fertilization must be alternated with meiosis in any regularly sexual life cycle. Thus regular fertilization from casual cannibalism probably evolved concurrently with meiotic cell division, which we examine in the next chapter.

There are at least as many mechanisms of recognition of mates as there are species of organisms, probably many more. Mate attraction and recognition evolved and reevolved literally millions of times in literally millions of ways. Specific examples of cell-surface interactions and compounds involved in mate attraction include proteins, steroids, isoprenoids, and others. Habitual mating was dependent on the ability of organisms to recognize specific mates.

Whatever mechanisms permitted cannibalism must have been strengthened by feedback to ritualize fusion and prevent mutual digestion. It is likely that destruction by digestion of one haploid cell by another is common in matings. Such destruction of mates by mistake, presumably a legacy from the days before fertilization became standardized, ought to be sought in studies of protoctists and other eukaryotes. From a philosophical-evolutionary standpoint it is interesting that at this time sex, rescue from death, cannibalism, and lack of digestion were much the same thing. Even today, in no population of mating organisms is 100 percent efficient mating achieved. The causes of failure to mate bear scrutiny: one imagines a trend through time of very low efficiency of mating in early protists to much higher efficiencies in later protoctists, plants, and animals.

The final step in the conversion of cannibalism to fertilization involves the regularization of karyogamy (the fusion of nuclei). We know that many organisms—all fungi, some red algae and foraminifera, and some ciliates—delay this step. Again the tendency must have been from low rates of nuclear fusion in early protoctists and fungi to very high rates in animals and plants. An examination of the physiology of karyogamy might well reveal the prerequisites and inhibitors of this step. No doubt different lineages will show different details of timing, chemistry, and control of mechanisms of nuclear fusion.

A strict series of events marking fertilization emerged in at least four ancient lineages: those ancestral to ciliates, animals, plants, and fungi. (Because the details of the mating-fertilization-meiosis cycle differ, fertilization probably evolved independently in many more than four lineages—perhaps in all those with meiotic sex, such as those labeled "m-f" for "meiosis-fertilization" in figure 22.) Each cannibalism or other production of a diploid cell had to be followed eventually by a reduction of chromosome number, that is, by a production of haploid cells. It is therefore likely that, in these four lineages at least, the transition from cannibalism to fertilization accompanied the transition from mitosis to meiosis. What specifically was required for the origin of meiotic reduction division from mitosis?

# 10 · PAIRING AND HALVING

## Fertilization and Meiosis

## DELAY OF REPLICATION OF MICROTUBULE ORGANIZING CENTERS LEADS TO MEIOSIS

Only diploid, tetraploid, or other even-number ploidy cells can divide by meiosis and survive. A diploid cell that divides by meiosis produces haploid cells, a tetraploid cell produces diploid cells. Any other known cell dividing by meiosis produces aneuploid offspring lacking a complete set of chromosomes and will tend to die. Diploidy is favored in the origin of meiosis because tetraploidy and other higher ploidies are more complex. More can go wrong and thus they are more prone to elimination by selection. Meiosis in many mastigotes and perhaps in the dinomastigote *Cryptothecodinium* occurs in only a single cell division that produces two haploid cells (Raikov, 1982). In other mastigotes, ciliates, and *Labyrinthula*, meiosis is a two-step process that involves two divisions (the second of which is mitotic) and produces four haploid cells. A diploid cell can divide only once by meiosis and survive. The resulting haploid cells must restore their chromosome numbers by incomplete cannibalism, fertilization, or another form of karyogamy.

Neither mitosis nor meiosis can be comprehended without an understanding of the behavior of microtubule organizing centers. The conversion of a mitotically dividing diploid cell to a meiotically dividing cell involves several steps. The most essential of these is a delay of replication of MTOCs attached to chromosomes (kinetochores) relative to other MTOCs. When the MTOCs attached to the chromosomes fail to replicate, the chromosomes to which they are attached fail to attach to the spindle. Entire chromosomes, rather than chromatids (half chromosomes), pass to the poles of the parental cell and thence, after cytokinesis, to the two offspring cells. A new round of chromatin synthesis fails to occur and the net effect is the production of cells with only one rather than two sets of chromosomes.

The pairing and subsequent segregation of homologous chromosomes without preliminary duplication, called "one-step meiosis," was observed by Cleveland in all the oxymonads studied (*Oxymonas, Saccinobaculus,* and *Notila*) as well as in *Leptospironympha* and *Urinympha,* two hypermastigotes. Crossing over, which requires double chromosomes, never occurs in organisms that have one-step meiosis (Raikov, 1982). The delay of kinetochore (MTOC) replication relative to cytokinesis is the central concept of the origin of meiosis from mitosis. Cleveland (1947) believed that meiosis evolved in protists from mitosis by these steps as a desperate effort to survive diploidy. The key factor in the origin of meiotic sexuality in haploid protists was the relief of diploidy. Haploid and other-ploid protists that had survived by merging as a response to starvation, desiccation, and such threats were now confronted with death by diploidy, tetraploidy, and other continued increases in ploidy. If the optimally adapted protist has been the original haploid form, those protists capable of relieving diploidy and returning to the haploid state enjoyed selective advantages and differentially reproduced when the environment returned to more generally favorable conditions. Cyclical environmental change is the rule whether the cycle is diurnal, lunar, annual, or another. Meiotic-fertilization cycles were probably as closely related to environmental cycles in the beginning as they are in so many animals today. The selection pressure to reduce diploidy and other higher ploidies was continuous and mechanisms for relief from diploidy ensued in such organisms. Yet desiccation and starvation, probably on a tidal, seasonal, or other cyclical basis, would encourage diploidy.

The ecology of protosexual environments most likely reinforced the evolution of meiotic sex by creating conditions alternately favoring haploid and diploid organisms. Though difficult to visualize, feedback cycles created among organisms comprising the environment may have been in large part responsible for such conditions. Often the products of fertilization are tough and hardy, more so than their haploid parents. For example, the zygospore of a large and successful group of fungi (phylum Zygomycota), such as blackbread molds *Rhizopus* or *Phycomyces,* is produced by an orgy of fertilizing nuclei. In these zygomycotes the hard-walled zygospore is a multinucleate diploid or other ploidy structure capable of withstanding desiccation and temperature extremes. The zygomycotes provide an example of alternating selection pressures in a life cycle, very characteristic of organisms in temperate climates. The diploid is the survival stage; it overwinters or survives the dry period. As soon as conditions adequate for growth return, the products of meiosis, haploid sporangiospores, germinate and grow.

Zygomycotes, although multinucleate, are fundamentally haploid, as presumably were all their ancestors. They revert to conjugation and diploidy only when threatened (fig. 36). In general, since they are smaller and require synthesis of only half the amount of DNA that diploids do, haploids are able to outgrow diploids of the same species. Haploids are probably able to adapt physiologically more quickly to changing environments than their diploid counterparts. Unfortunately, detailed comparisons of physiological adaptations of haploid and diploid individuals from the same parentage have not been made. Haploid yeast can, however, under optimal laboratory conditions, grow faster than diploid strains, so that at the end of a growth period the haploids are far more numerous than the diploids.

## SINE QUA NON OF MEIOSIS

The exchange of parts of chromosomes in the process of crossing over, the formation of the synaptonemal complex, and other special aspects of the meiotic process are known to occur in well-studied meioses like that which occurs in the testes of grasshoppers such as *Melanopus* or in the eggs of the horse worm *Ascaris*. The complete story of the origin of meiotic sex must of course explain these processes, yet they are considered to be embellishments on the essential aspects of meiosis and hence later developments. Beginning with a protist lamentably diploid, the absolutely minimal requirements for the appearance of meiosis as a mechanism to relieve this protist are two. First, the one round of kinetochore replication must fail, so that chromatids do not segregate—that is, they fail to move to opposite poles. Second, there must be a segregation of homologous chromosomes: the members of each pair of chromosomes must move to opposite poles.

The occurrence of these two events would have sufficed for complete relief from diploidy. The haploid protist, presumably now optimally surviving and dividing mitotically, would do so indefinitely. Eventually, however, the conditions forcing diploidy (starvation, thirst, etc.) would again select organisms capable of some sort of fusion. Whatever the cause, diploidy would need to recur before the protomeiotic events causing reversion to haploidy were repeated.

An example of alternating environmental pressures forcing the haploid-diploid cycle is seasonal change: even today winter conditions of cold or desiccation lead to the formation of dormant meiotic cysts. Presumably the quiescent diploid survives the pressures of inclement weather, whereas the actively metabolic haploid succumbs. Cyst formation is far more common in diploid than in haploid protists. In fact, dinomastigotes regularly fuse, two at a

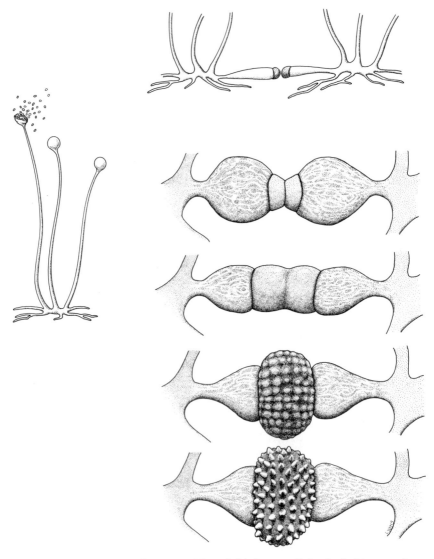

**Fig. 36.** Conjugation in *Rhizopus*. Although fundamentally haploid, *Rhizopus* forms a multinucleate, resistant structure (the so-called zygospore) when threatened. (Drawing by Christie Lyons.)

time, prior to encystment in a process that is not standard fertilization, and so do many species of protists (for example, chlorophytes) and fungi (zygomycotes). Contrariwise, in temperate weather the haploid forms are more active than the diploid. The zygospores of *Spirogyra* and *Chlamydomonas* fit this pattern, as do the zygotes of various zygomycotes. The well-protected zygotes (eggs) of *Hydra* provide a further example. Similarly, when nutrients are precipitately depleted in the water, dinomastigotes fuse and form resistant cysts. *Chlamydomonas* mate and form cysts when nitrogen is withheld from them.

The beginnings of the kinds of environmental pressures that alternately favored diploids and haploids were suggested by observations of the fate of the microbiota in the hindguts of wood-eating cockroaches and termites. The hypertrophied large intestine is sloughed off during moltings of the insect hosts. During these moltings the crowded inhabitants of the intestine are subjected to extremes: they lose contact with their food supply (wood eaten by the insect host) and are threatened by desiccation. Observing this periodic molting, Cleveland suggested that it placed selection pressure on, for example, the large wood-eating hypermastigote protists. During these times of stress the hungry protist bodies fused cannibalistically to form diploids.

Cleveland also observed already fused protists and interpreted their cytological behavior. He claimed that some underwent the events listed above: they failed to replicate their chromosomes. Others failed to replicate their "extranuclear organelles," in other words, the structures linking the chromosomes to the undulipodial bands that acted as kinetochores. He noted the segregation of entire chromosomes rather than chromatids to opposite poles of the cells. He tried very hard to convince his colleagues of the significance of these observations to the understanding of the origin of meiosis, but his entreaties fell on deaf ears. Because these events occurred in unfamiliar organisms with strange, rod-shaped "long centrioles," they were hard to link with familiar mammalian sex. Although we share Cleveland's difficulties in our attempts to explain the emergence of meiosis, recognition of the strong ancestral bonds among all eukaryotic cells makes sexual studies in protists more germane than ever before. Raikov (1982) has resuscitated interest in Cleveland's view on the origin of meiosis and has supported the soundness of the fundamental concepts with further data.

Cleveland was unable, of course, to observe each of the events listed above in a sequence in a single individual or set of individuals. Indeed, had he observed these phenomena in sequence, he would not have been able to distinguish the sequence from bona fide meiosis. He was mentally prepared to order the events, however, and so organized the information into a logical story in which all the cytological phenomena occurred in a regular orderly fashion. He further imagined that selection, alternatively on the haploid and diploid, would continue to refine these happenings. Eventually, orderly sequences of meiosis and fertilization would emerge. Many protist life cycles, such as those of some foraminifera, some heliozoans, some chlorophytes, and hypermastigotes, remained at the protist level. In lineages leading to our own ancestors, as we shall see, other events occurred that made the linkage between tissue differentiation and the requirement for meiosis obligatory. Before pursuing this peculiarity of our ancestry we must examine the likelihood that the most successful haploid-diploid cycles probably first evolved in protists with low chromosome numbers.

## MEIOSIS IN EARLY PROTISTS WITH LOW CHROMOSOME NUMBERS

We have seen that one of the essential steps in the origin of meiosis was the segregation of one member of each pair of chromosomes in the diploid cell for inclusion in an offspring cell. As Cleveland noted, the chances that a cell with a large number of chromosomes would (prior to the origin of the coordinating synaptonemal complex described in the next section) segregate homologues by chance alone was low. A diploid cell with few chromosomes is far more likely to produce a haploid offspring cell than is a diploid cell with many chromosomes. Table 11 shows chromosome combinations that can result

**Table 11.** Probabilities of a Diploid Cell Surviving Reduction Division

| Number of Chromosomes | Total Number of Differing Offspring | Chromosome Combinations* | Percent of Surviving Offspring |
|---|---|---|---|
| 2 | 3 | (1,1), (2,0), (0,2) | 33% |
| 4 | 5 | (2,2), (0,4), (1,3), (3,1), (4,0) | 20% |
| 8 | 9 | (4,4), (0,8), (1,7), (2,6), (3,5), (5,3), (6,2), (7,1), (8,0) | 11% |

*Only the euploid combinations (1,1), (2,2), and (4,4) will result in viable offspring. The other entries are lethal aneuploids.

from a dividing diploid cell. Assuming that aneuploids die and only haploids and diploids survive, the probability of survival is also shown. The conclusion is that either meiosis evolved in protists with low chromosome numbers or that immense numbers of protists with higher chromosome numbers perished during the refinement and control of the meiotic process. In fact, both most likely occurred. The probable polyphyly (multiple origins) of meiosis, coupled with the diversity and plurality of life, suggests that meiosis probably emerged in protoctists with varying numbers of chromosomes. It probably appeared in organisms with very few chromosomes and rarely if at all in organisms with very many chromosomes. The selection against aneuploids was relentless and totally successful, leading with time to elegant refinements of doubling and halving of chromosome numbers but only in certain protist lineages.

## ASSURANCE OF HOMOLOGUE SEGREGATION: THE SYNAPTONEMAL COMPLEX

Alternating selection pressures on the haploid and diploid parts of the life cycle must have led to increasingly refined mechanisms to protect against the formation of aneuploids and for their subsequent loss. The segregation of homologous chromosomes had to be better ensured than by the death of those failing to segregate properly. This kind of pressure led to the emergence of the synaptonemal complex, an elaborate structure, characteristic of the beginning of meiosis, that ties together homologous chromosomes in a diploid cell.

Synaptonemal complexes are structures of protein and RNA that lash together homologous chromosomes. They are frequent, indeed nearly universal, in the prophase of many meiotic divisions. But they are not limited to meiosis. In the tissue of the leaves of wheat plants, for example, synaptonemal complexes, situated at the membrane of the nucleus, are a constant feature (Stack and Brown, 1969). These cells are not, of course, destined to undergo meiosis; if they divide it is by standard mitosis.

It was suggested that the widespread tendency of homologues to pair, examples of which were listed by Stack and Brown (1969), is a preadaptation to the use of synaptonemal complexes in meiosis to ensure segregation of homologues. Why would homologous chromosomes in a diploid cell have found each other and bound to each other?

The tendency of homologous DNA to pair, as we have seen in chapter 4, is a pre-Phanerozoic legacy stemming from irreparable ultraviolet threats. The presence of homologous base sequences, presumably not only in chromatin but also in the genomes of the former spirochetes (that is, in the MTOCs), probably ensured pairing in many protists as a residuum from anaerobic days. One imagines that such pairing was utilized for repair of diploid cells in which one copy of the genome was damaged and the other was capable of complementation. Such repair must have occurred, and still occurs, independently of any meiotic process (Haskins and Therrien, 1978).

Those protists retaining the capacity for pairing and repair were preadapted for the appearance of the synaptonemal complex associated with meiosis. We know that at least several types of enzymes involved in carefully timed protein and RNA synthesis act during meiotic prophase I (Stern and Hotta, 1967). The entire repair capability (at least one seme) was probably retained as a unit in diploid cells and later put to use to restrain homologues on the metaphase plate in the first phases of the meiotic division. Such restraint by the elaborate axial structure, the synaptonemal complex formed between homologues, greatly increased the probability of segregation of these homologues, and hence of the meiotic production of euploidy by meiotic division. The testable aspect of this idea is the contention that in protists such as certain martiliads, paramyxea (Desportes, 1984), and the ciliates (Raikov, 1982) the presence of synaptonemal complexes will not invariably be associated with meiosis. Furthermore, such pairing may be observed in organisms, such as colonial euglenids and myxomycotes that lack meiosis entirely (Haskins and Therrien, 1978).

## CROSSING OVER

We have seen that extant meiosis generally involves both synaptonemal complexes and crossing over. We know of no organisms that display crossing over but fail to form synaptonemal complexes. The synaptonemal complex probably evolved first. Present as a manifestation of base pair homology in any diploid cell, the formations of the synaptonemal complex evolved in the transition from sloppy to regularized pairing of diploids. Once the synaptonemal complexes formed regularly between homologous chromosomes, and this evolution occurred only in certain limited lineages, it was a small step to reutilize the retained repair mechanisms—Archean legacies—for chromosome exchange and crossing over. Meiosis is still a protracted event; its different steps in the same individual (for example, a mammal) can today take hours, days, or even years. During their indirect and complex evolution protists must have suffered nucleic acid damages of various kinds. Some protists, capable of resuscitating the old ultraviolet repair systems, used their homologous chromosomes as templates. By definition these protists achieved crossing over. The search for the common ancestor of the host portion of the animal and plant cell should begin with the search for protist groups still enjoying the formation of synaptonemal complexes and crossing over. We suggest examination of diatoms, chytrids, or carefully selected sexual chlorophytes.

In any case it seems clear that meiosis was accomplished and flourished at first without the embellishment of the synaptonemal complex and crossing over. This still takes place in *Drosophila* males and the one-step meiotic protists discussed by Raikov (1982).

Sexuality in eukaryotes, including of course all protoctist sexuality, is of the meiotic sort. We believe it emerged from failure of cytokinesis after karyokinesis or from resistance to digestion in adult protists after cannibalistic attack by fellow members of their populations. Two adult nuclei, each with one set of chromosomes, fuse in the karyogamic process, each offering identical amounts of chromatin. In those lineages embellished by crossing over, DNA splicing and recombining enzymes are put to use. Note, however, that DNA recombination is not intrinsic to the meiotic sexual process. In meiotic sex recombination is on the level of cell and nuclear fusion. Fertilizations bring chromosomes of different parental nuclei into a common nucleus, and the DNA of chromosomes does not necessarily recombine. In refinements of the meiotic process, such as the transfer of nuclei in fungi and conjugating protists, the recombination is on the nuclear level only (that is, karyogamy occurs without syngamy of the rest of the cell). Among two eukaryotes, however, the minimal sexual event involves the introduction of two nuclei into a common cytoplasm.

MTOCs are associated not only with kinetochores of each chromosome, but often with "intranuclear division centers" or "nucleus-associated organelles" (NAOs; Girbardt and Hadrich, 1975; Heath, 1980a, 1980b) as well. MTOCs thus travel with their associated nuclei in the karyogamic process. We can now interpret major meiotic events in the light of the spirochete hypothesis. At every fusion, whether cannibalistic or genuinely fertilizing, the resultant zygote nucleus receives what is equivalent to a new supply of once-foreign genomes able to replicate, to produce certain gene products, and, most importantly, to confer intricate internal movements on their host cells. The active participants in this MTOC communication system can be thought of as "spirochetal secret agents" (Margulis and Sagan, n.d.). We believe the transfer of the genetic potential of the original microbes to each newly formed diploid cell was and still is crucial for the origin of differentiated tissues in the body cells of ancestral and extant animals. Generations of would-be cannibal haploids, protists interested merely in living through the dry season or the winter, passed around their incessantly motile MTOC heritage. This heritage of the spirochetes is now restricted in its replication. MTOCs must replicate using the autopoietic systems of the host protist cells with which they fused. But the MTOCs still maintain their original immense potential for various kinds of microbial motility. The retention of undulipodial locomotion was crucial for the competitive heterotrophic ancestors to animals. The animal ancestors paid a price for the retention and exchange of the well-developed motile systems of their benign internal parasites, their spirochete-undulipodia. What was this price?

In our lineage once a host protist committed its MTOC to intracellular motility by undulipodia, that host protist cell could no longer divide by mitosis. To retain undulipodial motility *and* mitosis, such a host cell was required to reproduce, to double, to become two host cells inextricably joined. From these frustrating origins—the mutual exclusion of protruding motility organelles and mitosis—arose animal-style differentiation.

Theoretically, the first step on the staircase of irreversible MTOC specialization in animal cells began with the attachment of spirochetes to host cells, long before animal cells had diverged from the main eukaryotic line. Today spirochete bacteria live in the same environments as protists. In termites they swim alongside each other, often in synchrony (fig. 37E). Spirochetes have been seen to attach in large numbers to protists such as pyrsonymphids and dienymphids (fig. 37A, B, D). Feeding at the periphery of their hosts, modern spirochetes have been observed to form both casual or highly organized attachment sites. Live spirochetes have also been seen invading and living inside protist hosts (fig. 37C). *Pillotina* spirochetes, for example, have been seen to enter the parabasalid *Trichonympha*, a large wood-eating microbe that lives in the termite hindgut. Elaborate attachment sites of spirochetes to their hosts also have become established. This is the case today with *Pyrsonympha* and *Mixotricha paradoxa*, both of which are

**Fig. 37.** Spirochetes and their attachments. A. *Pillotina* spirochete (P) and *Hollandina* spirochete (H) along with several undulipodia (u) from protists in the termite gut community. Transverse sections, transmission electron microscopy (bar = 0.5 micrometers). (Courtesy of Dr. David Chase.) B. Spirochetes, from termite intestine, attached to an unidentified host (bar = 10 micrometers). C. Electron micrograph (bar = 1 micrometer) of spirochete (*Pillotina* sp., labeled P) inside host protist seen in longitudinal (l) and transverse (t) section. D. Electron micrograph (bar = 1 micrometer) of a spirochete attachment. E. Light micrograph (bar = 10 micrometers) of spirochetes, not physically attached, moving in synchrony.

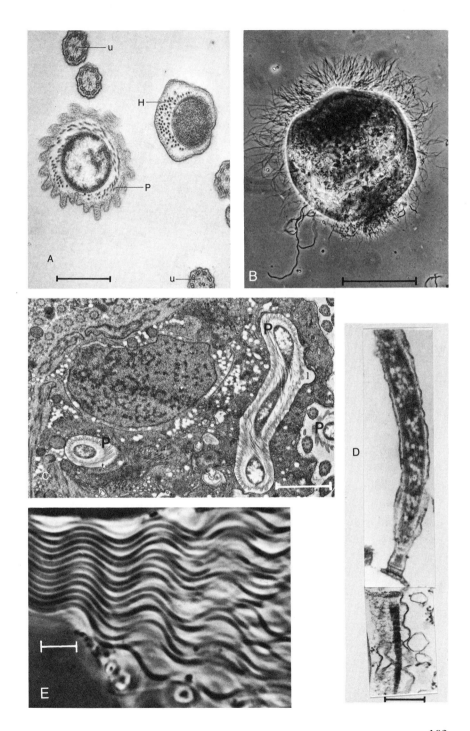

termite-hindgut microbes with distinct internal and external bacterial symbionts (fig. 37D).

Spirochetes were whittled down to their MTOC essence by the forces of natural selection in a biotic environment. Theoretically they evolved from surface symbionts to specialized structures, such as cilia, and "chromosome movers," such as the mitotic spindle microtubules. Although the extent of the symbiotic merger here may be unprecedented in its pervasiveness and intimacy, the differentiation of spirochetes into subcellular components is understandable by analogy with other symbionts. Cryptomonad symbionts in the ciliate *Mesodinium rubrum* show the type of differentiation (dedifferentiation, or a "reverse" differentiation to a simpler form) we suspect to have taken place in the spirochetal ancestors to MTOCs. The cryptomonad symbiont undergoes dedifferentiation when it enters into its ciliate hosts. Once inside, the plastids separate from the nucleus, ending up at different locations in their host cells. Sometimes the plastids and mitochondria of the former cryptomonad reproduce faster than its nucleus. This leads to different ratios of nuclei and plastids than were originally present in the invading cryptomonad. Symbiont organellar remains become situated in different portions of the host cell (Taylor, 1983).

The absence of DNA in centrioles, kinetosomes, or MTOCs is sometimes invoked as an argument against the possibility that these organelles originated from spirochetes. But two observations may be taken as precedents for the loss of DNA from the organelles and its transfer to the nucleus. That MTOCs lack DNA inside them does not rule out the idea that they are legacies of extremely altered, streamlined symbiotic genomes that have transferred their DNA to elsewhere in the cell. It does not preclude the possibility that they store their information in RNA.

Symbiontlike gamma particles in *Blastocladiella* may help to shed some light on the mysterious nature of MTOCs. These elusive, bacteria-sized, membranous particles are the sources of the enzyme chitin-synthetase in these protoctists. Gamma particles contain DNA and RNA but lack ribosomes. They appear as membrane-bounded entities only at certain times in the *Blastocladiella* life cycle. The rest of the time they are not even visible—as is the case with MTOCs.

Another observation is relevant to the absence of DNA in MTOCs. The presence of DNA in a symbiont does not appear to be absolutely required once the symbiont has been incorporated into a living environment such as that of the eukaryotic cell. Particularly fascinating in this regard has been the discovery of chloroplasts totally lacking DNA in the protist *Acetabularia* (Woodstock and Bogorad, 1970). Since the majority of biologists studying cell evolution believe chloroplasts to be symbiotically derived, this observation means that so long as one complete copy of organellar DNA is present in the cell, wherever it may be, the DNA of other organelles may safely be lost. The spirochete remnants seem to have differentiated to RNP (ribonucleoprotein) in many kinds of cell morphogeneses, most of which involved assembly of microtubule protein into microtubules. Apparently autopoiesis and reproduction of spirochetal remnants in the form of MTOCs no longer depends on their direct association with DNA. The remnant DNA originally and still responsible for the existence of the MTOCs has been transferred to the nucleus or lost through transfer to RNA. Some of the information for cell morphogenesis is assumed to be retained by RNA. Some of the original spirochete DNA (for example, that coding for tubulin protein) must have entered the nucleus and been incorporated into chromatin.

The critical genetic work of S. Dutcher, D. Luck, and their colleagues promises to resolve the issue of the genetic status of the MTOCs of kinetosomes (Huang et al., 1982). Positional information in both ciliates (Martin, 1984) and *Chlamydomonas* (Dutcher, 1984) is inherited independently of the nucleus, mitochondria, and plastid genes. In *Chlamydomonas* a tightly linked cluster of genes determining ultrastructural information of kinetosomes may be carried on the kinetosomal RNA (Dutcher, 1984). By hypothesis this particular kinetosomal RNA is homologous to spirochetal DNA and *is* the "spirochetal secret agent."

In all stable symbioses the symbionts and hosts eventually match reproductive rates. Regulation of symbiont reproduction rate is well known in *Paramecium bursaria*, a ciliate with a large (from thirty to several hundred) internal population of photosynthetic symbionts (*Chlorella* chlorophytes). Many strains of *Hydra viridis*, coelenterate animals that are green because they also harbor symbiotic green algae, are known from Europe, North America, and Japan. The hydra symbionts, also chlorophytes of the genus *Chlorella*, reproduce in concert with their host. A stable ratio of *Chlorella* cells to hydra host cells is maintained. Subject to environmental perturbation such as changes in light intensity, starvation, cold, and inhibitors of photosynthesis (such as the inhibitor of oxygen-producing photosynthesis, dichloryl methyl urea, or DCMU), the ratios of symbionts to host change in predictable ways. In one study, for example, starvation in the dark increased the number of algae from 12 to 27 per hydra cell. Within a day the normal number of about 12 algae per hydra digestive cell was restored (Muscatine and Neckelmann, 1983).

The coordination of reproductive rates in host-symbiont relationships, subject to rigorous natural selection, is intermediary in evolutionary time between spirochete invasions and the establishment of a subtle network of MTOCs. By hypothesis, spirochetes eventually deployed their parts to various fixed regions inside their hosts. Those parts retaining nucleic acids became "spirochetal remnants." Spirochetal remnants include the 9 + 0 centriole structure and the 9 + 0 kinetosome, required structures for the infrastructure of all undulipodia. Morphogenetic MTOCs, such as "division centers" in heliozoans and heliomastigotes (helioflagellates; Brugerolle and Mignot, 1984) and kinetochores and pericentriolar material in mammalian cells (Calarco-Gillam et al., 1983), share what is probably the spirochetal ancestry. Spirochetal remnants in the form of ciliate cortical MTOCs determine taxon-specific kinetids. Their retention led to the profound cortical, pattern-based speciation of ciliates (Corliss, 1979; Lynn and Small, 1985; n.d.).

A central aspect of our thesis is that spirochetal remnants, in the form of MTOCs, evolved in different ways in different protoctist lineages. It was a peculiarity of the animal cell lineage, multicellular as it was, that once any spirochetal remnant inside a given cell was committed to produce an undulipodium the cell itself could never again divide by mitosis. In animal lineages undulipodiated cells never divide. In certain protoctist lineages, however, this is not necessarily the case. Some dinomastigotes and some amoebomastigotes, such as *Naeglaria* and probably *Paratetramitus*, regularly withdraw and then regrow their undulipodia (fig. 38). While undulipodiated they cannot reproduce. Some chytrids, such as *Blastocladiella*, also alternate between the mutually exclusive processes of cell division and undulipodia production. *Naeglaria* has two undulipodia, which it uses to swim about its watery environment. *Blastocladiella* has one. But in both cases the undulipodia are retracted prior to cell division. *Paratetramitus jugosus* (an amoebomastigote similar to *Naeglaria*) probably resorbs its undulipodia prior to reproduction as well (fig. 39). It appears that the MTOCs used to produce undulipodia may be co-opted in some way to the task of mitosis.

This trap of differentiation was never overcome in individual cells of animal lineages: we live with the impatience of our motile cell inhabitants. Any given spirochetal remnant, as MTOC, may either produce an undulipodium or grow microtubules to segregate chromatin in mitotic cell division. In animal lineages the individual MTOC may never do both at once. From this internal commitment of the former bacteria inhabiting our ancestors, we believe, came the imperative of elaborate cell differentiation.

Multiple cells performed multiple tasks. The connection is obligate: it is intrinsic to our very being. Tissue differentiation, the process that makes us animals with all of our restlessness, is coupled inextricably to the retention of undifferentiated MTOCs for genetic continuity. In what cells were MTOCs

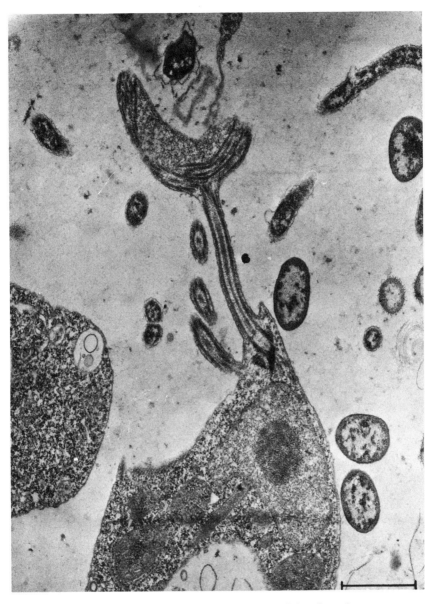

**Fig. 38.** *Paratetramitus* sp. apparently resorbing its undulipodium (bar = 1 micrometer). (Courtesy of Floyd O. Craft, Boston University.)

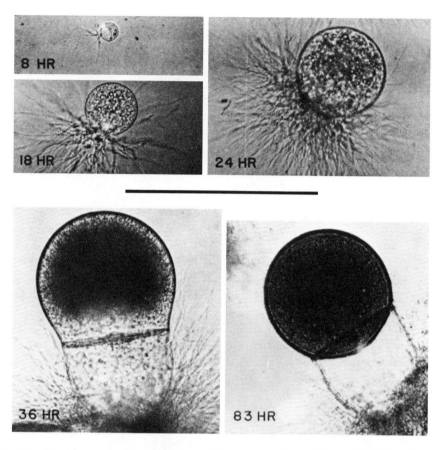

**Fig. 39.** Development of the ordinary colorless sporangium of *Blastocladiella emersonii*. The hours are time elapsed after water was added to an initial small dry sporangium. After 18 hours, there is swelling and a proliferation of rhizoids. After 36 hours, the protoplasm has migrated into the anterior cell that becomes the sporangium. After 83 hours, the sporangium has thickened, and zoospores have begun to differentiate from the coenocytic nuclei inside. Light micrograph (bar = 1 micrometer). (Courtesy of E. C. Cantino.)

capable of reproduction reserved? In what cells was the capacity for mitosis retained? The retention of functional, undifferentiated MTOCs always capable of entering mitosis occurs in the cells of the germ line alone. These cells, sperm, eggs, and spores are destined to repeat the incomplete cannibalistic encounters of their ancestors.

As soon as differentiation emerged in organisms committed to cannibalistic fusions and meiotic reductions, animal-style multicellular individuals evolved. At first they were composed of at least two cells: one was committed to undulipodia-producing MTOCs and therefore unable to divide by mitosis; the other reserved uncommitted MTOCs, incapable of moving by undulipodia but capable of continued division by mitosis.

In the earliest animals reproduction became enslaved, imprisoned by sexuality. Many attempts to escape this imprisonment ensued, as we shall see. In some lineages meiotic sexuality was formally bypassed but it never disappeared entirely. In all ancestral protoctist lineages, as long as there was any sort of tissue differentiation, fertilization-meiosis sexual cycles were retained. The essence of animal-style differentiation is the formation of cells capable of nonmitotic internal motility (such as the growth of dendrites and axons in nerve cells, melanocytes in pigmented tissue, and cilia in epithelium). All these cells simultaneously lose the ability to divide by mitosis. From the beginning the regularization of fertilization in this lineage involved the fusion of a cell with uncommitted MTOCs (eggs) with a committed, motile, undulipodiated cell (sperm). The perpetuation of the cells with the uncommitted MTOCs required this fusion. To retain differentiation in our ancestral animal lineage was to retain meiotic sexuality.

The coupling of sex with reproduction in animals came from inauspicious protist beginnings. Animal-style meiotic sex is associated with a product of a particular sort of anisogamy. Isogamy is the fusion of nuclei from morphologically identical partners. From isogamy in the aborted cannibalistic-fertilization events came anisogamy: the fusion of big eggs and small sperm. Comparable trends from isogamy to anisogamy occurred in many protists and their protoctist descendants. Most of these conversions did not involve differentiation and obligate sexuality. This again makes us aware of the peculiarity of our animal ancestry.

# 11 • MEIOSIS AND CELL DIFFERENTIATION

## From Microbial Community Ecology to Endocytobiology

### THE ORIGINAL EUKARYOTES AS MICROBIAL COMMUNITIES

A central thesis of this book is that the eukaryotic cell is homologous to a community of microorganisms. Eukaryotic cells are not simply larger, more complex prokaryotic cells. We recognize cell biology to be endocytobiology: the study of several kinds of cells within cells (Schenck and Schwemmler, 1983). Cell biology becomes the study of tightly integrated microbial symbionts. We see in this chapter how our perspective changes the way we view embryogenesis, the origin of embryos. Development and differentiation from the symbiotic vantage point are not singular processes but rigidly controlled mechanisms of microbial community ecology.

Cell differentiation in animals and plants originated from the population dynamics and interaction of interdependent members of former bacterial communities. As in any community that grows and alters, the numbers of community members and their ratios change through time. Each autopoietic member of the community itself arises by reproduction and grows by the synthesis of nucleic acids, proteins, carbohydrates, and lipids. In the case of the major heterotrophic eukaryotes, protoctists, fungi, and animals, at least three original bacterial components can be counted. These members of the communities comprising the eukaryotic cell are nucleocytoplasmic, mitochondrial, and undulipodial entities. In the photosynthetic eukaryotes, algae and plants, the communities comprising the cell include a fourth autopoietic entity: the genome and protein synthetic system of plastids.

Each of the original bacteria systems began with both a genome and the protein synthetic machinery required to express that genome. Regardless of the position in the cell each bacterial system occupies, the minimal genome

170

and protein synthetic machinery of autopoiesis should still be present at all times. Because of the necessarily conservative nature of autopoiesis, each of the several cell genomes was retained. It is clear, however, that the original genetic systems of the microbial components that make up the eukaryotic cell of today have been profoundly modified by long and intimate association with each other.

There are some hints already that specialization of cells as they differentiate for given functions is the outward evidence for changes in ratios of the different genetic components that make up the cells. Our best example derives from an obscure pond-water organism. This protoctist, *Blastocladiella*, has provided us with examples, at least in principle, of the phenomenon of differentiation as a manifestation of differing ratios of genetic components of cells. Sometimes identified as plants, sometimes fungi, the group of multicellular, differentiated, undulipodiated, eukaryotic organisms to which *Blastocladiella* belongs are members of the chytrid phylum of the protoctist kingdom in the five-kingdom classification (Margulis and Schwartz, 1982). Perhaps because of the confusion surrounding their classification, the elegant research work on them has been for the most part ignored by mainstream students of differentiation phenomena. Nevertheless, the fascinating discoveries of Edward Cantino at Michigan State University concerning *Blastocladiella* and its gamma particles are models for understanding phenomena of cell-level differentiation.

*Blastocladiella* is probably best thought of as a water mold that develops from mastigotes: undulipodiated, spermlike single cells. These cells, called "zoospores," are the only means of reproduction the organism possesses. But a single zoospore enjoys at least three different sorts of developmental pathways, depending on the conditions in which it finds itself.

Starvation restricts zoospore development such that only a microscopic chytrid body, just slightly larger than the zoospore itself, is formed. High concentrations of carbon dioxide, however, induce a different sort of development in *Blastocladiella*. Prompted by the suffocating quantities of carbon dioxide that are characteristic of crowded conditions, *Blastocladiella* forms a relatively monstrous structure, many times the size of a single zoospore, called an "ordinary colorless thallus." Still another developmental pathway occurs when there is a lack of sufficient moisture. Desiccation induces the zoospore to make a brown, resistant structure called a "resistant sporangium" even larger than the ordinary colorless thallus. Hence three sorts of structures—resistant sporangia, ordinary colorless thalli, and tiny chytrids—may be developed. All these variations on the *Blastocladiella* life cycle differentiate from zoospores that resorb their undulipodia (see fig. 39). The cell walls in all these developmental stages are composed of chitin, a hard, polysaccharide material that is made up of nitrogenous sugars.

Why does a single zoospore respond by differentiating three very different sorts of bodies depending on the cues from the environment? The answer to this question has several levels. From the point of view of selection, those zoospores capable of such rapid differentiation, a differentiation that responds to the needs of the crowded, dry, or nutrient-poor environment, leave more offspring. But Cantino and his colleagues have been able to recognize another correlation: that between the fate of the zoospore and the number of gamma particles it contains.

Gamma particles contain DNA and messenger RNA. They are cytoplasmic organelles about one micron across that display a life cycle of their own. They are the sites, in the cell, of chitin-synthetase and other proteins. Chitin-synthetase is the enzyme that forms the nitrogen-rich, celluloselike polymer, chitin. The large, resistant, thick-walled sporangia naturally contain far more chitin than the zoospore. The ordinary colorless thallus wall contains more chitin than the tiny chytrids and less than the resistant sporangium.

The origin of gamma particles is obscure; they may have begun as viruses (Cantino and Mills, 1979). Still more likely, in our view, is the possibility that they originated as symbiotic bacteria.

The relationship between the gamma particles and differentiation of *Blastocladiella* is remarkable. Gamma particles reproduce, but not by direct division, inside the cytoplasm of zoospores. (One sees an increase in number of gamma particles, but the details of reproduction are not clear because of the difficulty in observing gamma-particle division stages. No direct bacteria-style cell division has been seen.) Later the gamma particles move to a new place of residence along the surfaces of the zoospore nuclei. Those zoospores that contain only one or two gamma particles produce the tiny chytrid bodies; those that contain some eight to ten form ordinary colorless thalli. The zoospores that contain sixteen to thirty-two gamma particles form the thick-walled resistant sporangia. Here, in *Blastocladiella*, the number of gamma particles relative to the host nucleus determines the differentiated fate of the zoospore. Each zoospore is totipotent in the sense that, if conditions change, its future will change accordingly.

Again we see that changes in number of reproducing entities lead to qualitative changes in morphology and function. Gamma particles seem to align themselves alongside the nucleus at certain stages. They do not seem to contain their own ribosomes, even though they do have the required messenger RNA for production of chitin-synthetase. The absence of ribosomes inside the gamma particle suggests that the host cell's ribosomes are used for synthesis of the gamma particle's product. Whether gamma particles began as bacteria that have lost ribosomes or as viruses that never had them is not known (Cantino and Mills, 1983). In any case the principle of differentiation developed in this book is beautifully illustrated by the example of *Blastocladiella:* the ratio of self-replicating entities comprising the cell determines the differentiated fate of that cell. Eukaryotic cells are communities that have learned to play a biological "numbers game."

The name for the genetic status of the eukaryotic cell as microbial community, composed of several genomes from different sources, is *heterogenomic.* Techniques of molecular biology have already established the heterogenomic nature of eukaryotes. The nucleocytoplasm, mitochondria, and plastids are genetic systems with distinct microbial ancestry (Gray, 1983). As we have seen, the spirochetal ancestry of the undulipodia has not yet been demonstrated, but progress has been made. Microtubule-like protein has been identified in spirochetes (Fracek, 1984; Obar, 1985). RNA synthesis accompanies the genesis of new kinetosomes (Hartman, Puma, and Gurney, 1974; Younger et al., 1972; Dippell, 1976). This RNA is functional in the microtubule organizing process (Heidemann, Sander, and Kirschner, 1977). Kinetosomal RNA may genetically determine the ultrastructure of *Chlamydomonas* kinetosomes (Dutcher, 1984). Yet there still is a paucity of the molecular biological evidence needed to prove the spirochetal connection (see chapter 6). Nevertheless it is mandatory for our analysis of the origin of meiotic sex that we accept the genetic status of all eukaryotic cells as coevolved microbial communities. Without this concept, in our opinion, there can be no understanding of the origin and evolution of development and differentiation.

## ANALOGIES WITH ECOLOGY: THE CELL AS COMMUNITY

In most animals and plants fertilization and meiosis regularly intervene in the sexual cycle. The individual animal or plant returns each generation to a single cell: a zygote with organelles such as mitochondria and plastids derived from the female parent. Finely tuned biological control systems must be required to ensure the return of the diploid cell to two copies of the nuclear genes and at least one copy each of the mitochondria and plastid genes. The evolution of such control systems was undoubtedly complex and will be revealed only as the control systems themselves come to light. In the meantime it may be useful to pursue the nature of such control systems by developing analogies with known complex communities. Unfortunately, the cell as a microbial community is such a basic module of living systems, so intrinsic and complex, that nearly any analogy will seem lackluster.

We may, however, compare the meiotic return, say, of mammals in each generation, to a set number of copies of each gene with the predictable, regulated control of ecological succession in the case of a northern temperate forest. The return of a climax forest of given dimensions first to an old field and then back again to the climax forest of these same dimensions bears a faint resemblance to the control process of meiosis. Though equally poor in the precision and integration needed to evoke the coordination of the eukaryotic cell, other large-scale ecological analogies present themselves. To be analogous to the community ecology of the eukaryotic individual, the fire-stimulated chaparral would have to begin its cycle with a given volume, a given number of community members in specific positions. Perhaps stimulated by periodic burning, the chaparral would have to behave with predictable regularity. The spruce bog successional cycle, which begins with newly excavated glacial lakes, proceeds from communities of low-lying ground cover, sphagnum moss, and herbaceous plants such as *Vaccinia* to *Alnus* and *Betula* shrubs and then to spruce, pine, and other conifers—again all with predictable regularity. The cyclical precision of communities as large as forests is far less fixed and less predictable than that of the microbial communities that we recognize as embryos.

Community development, on the scale of spruce bog and chaparral forest, is the complex result of interacting numbers and kinds of organisms responding to environmental changes, many of which the growing population of organisms induce themselves. These large-scale ecological cycles, although not at all random, lack by far the precision and predictability associated with individual animal and plant development (ontogeny). Few mechanisms exist that limit the spatial distribution and ensure the orderly appearance and the control of the populations—numbers of members of species—on the scale of lakes or forests.

Natural selection simply has not been operating as long or as powerfully on the larger ecological communities. Vastly fewer organisms (trees) living in a medium that is more confining (the forest floor) have rarely been exposed to selection pressures that prompt, as a matter of survival, the level of genetic, metabolic, and behavioral interaction that is standard in bacteria. The huge, heterogenomic, forest ecological community is at a larger scale of self-organization than the microscopic, heterogenomic, ecological community we call the cell. At the forest end of the scale, trees and mammals, shrubs and birds survive with their fellow community members in a relatively loose matrix. This may change. As natural selection has time to beleaguer and sift through the organisms of the macrocosm, superorganisms (precisely integrated ecological communities) may emerge that embody a level of coordination and complexity analogous to that found in the eukaryotic cell components of plant and animal bodies. For forest-level ecological succession to be repeated in a highly heritable way, genetic, metabolic, and behavioral complexity would need to be ensured by elaborate control mechanisms. Such forest-level control mechanisms would be analogous to those of meiosis—though no doubt radically different in detail.

Perhaps the communities most approaching cyclical, controlled patterns of component population growth are those of the hundguts of termites and wood-eating cockroaches (To, 1978; Margulis, Chase, and Guerrero, 1986). The repetitive nature of the development of these communities is imposed by the development and behavior (the periodic molting) of their insect hosts.

The microbial communities that become the protoctist cells in which meiosis evolved are more ancient, more complex, and more tightly regulated than either forest communities or termite microcosms. Cells of protoctists, fungi, plants, and animals have evolved precise mechanisms to ensure specific, repeating population cycles. The most crucial requisite of community control is the reservation of at least one complete set of genomes (three sets of heterologous genomes in heterotrophic eukaryotes and four sets in photosynthetic eukaryotes) confined within a plasma membrane throughout the life cycle of all eukaryotes. This entity is capable of indeterminate growth: it is immortal.

The retention of the master set of genomes in animals has been recognized as maintaining the *germ plasm* since the work of August Weismann in the nineteenth century. Weismann undoubtedly underestimated the wide capacity of many animals and nearly all plants to retain the master set of community genomes outside the germ line. (The germ line, especially as studied in vertebrate animals, is the set of cells destined to become gametes, such as eggs and sperm.) Thus most species of eukaryotes can form ramets (clones, asexual individuals capable of further growth through organs such as the gemma of sponges or the budded hydroids of coelenterates; see Buss, 1983a). The soma, or body tissue, and its changes are probably far more important for the evolutionary process than hitherto imagined (Buss, n.d.). Nevertheless the fundamental constraint of development, brilliantly conceived of by Weismann, is the obligatory nature of the retention throughout development of the entire set of genetic possibilities in at least one cell lineage in good working order and capable of continued growth and cell reproduction. With the knowledge of the genetic status of cell organelles we must expand Weismann's concept of germ plasm to include at least one set of "cytoplasmic" genes, that is, genes for each of the reproducing organelles. All eukaryotic individuals must reserve, in a form capable of continued reproduction, their genetic components, the remnant bacteria in the combined form of the nucleocytoplasmic, mitochondrial, plastid, and undulipodial genomes. If we accept the cell as a microbial community, the germ plasm is equivalent to component autopoiesis: a complete set of heterologous genomes and their protein synthetic systems contained within a membranous package—not the nuclear membrane but the plasma membrane.

We can apply principles of community ecology directly to the development of the individual. Just as each population of a large community can grow to some extent independently of its fellow community members, each of the different members of the cell population can *hypertrophy* (grow more numerous than one or another of its cell associates, as do the mitochondria in relation to the nucleocytoplasm of insect muscle) or *hypotrophy* (decrease in number relative to its associates, as in loss of undulipodia in mature suctorians relative to the nucleocytoplasm). Indeed, various organelles can be lost entirely during differentiation. This occurs, for example, in the production of enucleated red blood cells in mammals or in the dedifferentiation of mitochondria in yeast grown in fermentation media. During the course of differentiation the numbers and products of the heterologous genomes change, just as the numbers and qualities of the members of communities change through time in any natural setting. In order for the genetic potential to be retained in any community, however, at least one copy of each heterologous genome must be retained.

Our central hypothesis is that there exists in all sexual eukaryotes a formalized, regulated method by which at least one copy of each heterologous genome, with at least one copy of each gene, is maintained. (Here the analogy with large communities clearly fails: large communities have not yet developed cyclical methods to retain, in a given volume, a copy of each heterologous genome and its protein synthesizing system.) The cyclical mechanism by which morphologically complex protoctists, fungi, animals, and plants ensure this retention of at least one copy of each gene in each heterologous genome is meiosis, which must be followed by fertilization to ensure the full cycle.

## MEIOSIS IS REQUIRED FOR DIFFERENTIATION AND DEVELOPMENT

All regulated systems, from the most simply engineered thermostat to large computers, have built-in error-correcting processes. The thermometer of a thermostat rises, sending a signal to shut off the furnace. When the room consequently cools, the lowering of the thermometer switches the furnace back on. For any system to achieve stability, even stability that changes through time, errors must be sensed and corrected. These are not errors in the sense of mistakes but merely perturbations. To be sensed and corrected, values of these errors must be compared with some standard values. The standard basis for comparison in the informational systems of engineering is called "the set point." A homeostatic system is regulated around "set points." If the system and its standard values, or set points, are perceptibly changing with time, the system is homeorrhetic instead of homeostatic (Waddington, 1976). The set points of homeorrhetic systems are called "operating points." On these general principles alone it is inferred that eukaryotes have self-regulating systems at the level of the molecular biology of the cell that ensure the inclusion of the necessary "starting genes" in each generation. Although we compare the homeostatic activities of eukaryotes beginning to undergo differentiation with the error correction of simple machines, we do so only for simplicity's sake. In fact, the reader should keep in mind that the self-maintaining processes of organisms are far more complicated than even the most sophisticated computers.

Although there is no direct evidence yet for some sort of meiotic cell control process, the need for at least a nucleic acid editing process has been inferred. The paucity of evidence probably comes from the failure to collect molecular biological data organized in search of meiotic control functions.

Eukaryotic organisms have far more DNA than they need to code directly for their protein synthesis. They contain, in fact, from ten to about a thousand times the DNA they require to make proteins, depending on species. In any given eukaryotic cell at any given time, the quantity of DNA in each of the classes of organelles (nuclei, plastids, mitochondria) varies depending on many factors. For example, quantities of DNA vary in different types of tissue and at different ages of the cell or organism. The organelles themselves are present in variable quantities per cell. Yet we know that, for a given tissue, a cell or organism of a given age will contain a more or less measured amount of DNA.

Development proceeds in an orderly fashion, each generation assuring an approximate constancy of DNA, organelles, cells, and tissues. Some assessment and correction or controlling processes must exist. Viewed from our standpoint, the zygote is a microbial community composed of three or four different types, and the developing individual is the external manifestation of the growth and interaction of populations of these types. We have inferred that cyclical population succession, which is equivalent to accurate development, requires error correction. Checks must be made on the members of the community to assure that they and their genetic parts necessary for continued protein synthesis are all present and accounted for. In order to begin again in each generation, the development of an individual as complex as a brown alga, a shelf fungus, or a sea urchin must include some verification that the components—the former symbionts, each with its genes and protein synthetic systems—are intact. The minimal system with its components, genes, and whatever else is required for these genes to express themselves must be present at the beginning of each life cycle. Biochemical feedback systems must ensure that at least one copy of each gene, organized into sets of chromosomes and organellar genomes, is present at the beginning of each new generation of individuals. If this hypothesis is correct meiotic sexual cycles may be required as an error-correcting process, a sort of bio-inventory checking and adjusting the mechanisms necessary for differentiation and development. Meiosis, evolved from the mitotic "dance," now functions as a choreographer-producer, an organizer that verifies and directs the diverse components of the eukaryotic life cycle. That it is difficult to distinguish the dancers from the dance only means that more work needs to be done.

Natural selection acts on the developing differentiating protoctist, animal, plant, or fungus. Sexuality thus has been obligately correlated with existence in morphologically complex eukaryotes. Loss of meiosis, often associated with loss of sexuality, is generally lethal in such organisms.

It is certain, at any rate, that those eukaryotes that demonstrate the phenomenon of differentiation also undergo meiosis. Differentiation, whether on the level of the cell (as, for example, in foraminifera or the ciliates) or on the level of the individual multicellular organism (protoctist, animal, plant, or fungus), is accompanied by meiosis and fertilization at some point during the life cycle. Natural selection to bypass meiotic sexuality has occurred many times in evolution, for example, in the appearance of parthenogenetic animals and apomictic (formerly sexual) plants (fig. 40). Such selection has resulted in many circuitous and intricate developmental life cycles in members of all four of the eukaryotic kingdoms. But time and again the entire loss of meiotic sexuality has not been tolerated. The loss of meiosis without a trace of its former existence, like the complete loss of any seme, is simply not plausible. If evolution boasts any rules at all one of them is that semes, multigenic traits with identifiable selective advantage, are never entirely lost by eukaryotes.

## FROM INCOMPLETE CANNIBALISM TO MORPHOLOGICALLY COMPLEX INDIVIDUALS

It is folly to endow the evolutionary process with foresight. Any explanation of the rise of developmentally complex sexual eukaryotes must provide steps that imply no planning.

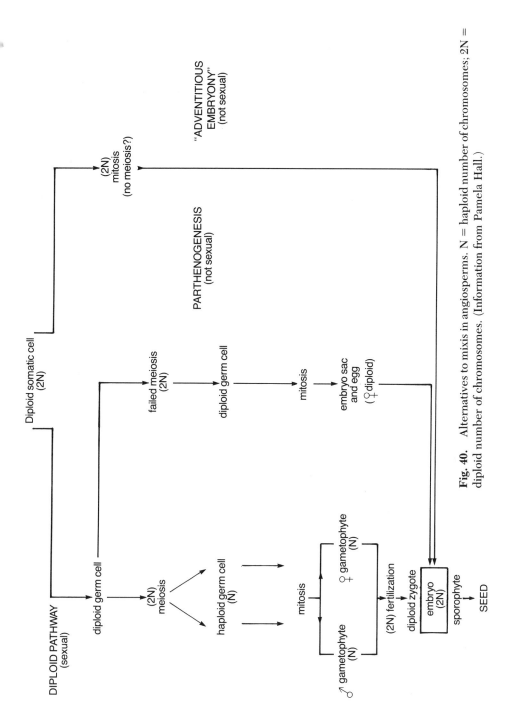

**Fig. 40.** Alternatives to mixis in angiosperms. N = haploid number of chromosomes; 2N = diploid number of chromosomes. (Information from Pamela Hall.)

179

The first protoctists on the lineage leading to animals and plants were two-celled: one cell was capable of mitotic division and the other, equipped with undulipodia, of movement. In the inimitable fashion of expanding autopoietic systems, organisms in this lineage must have succumbed to the temptation to exercise their mitotic option: they grew larger and many-celled by mitotic division. They retained in one mass the progeny of these mitotic divisions. In many such masses the MTOCs differentiated, giving up their capability for reproduction. They formed undulipodia and multiciliated cell layers like those found in the most simply organized of all animals alive today, *Trichoplax adhaerens*. This relic of the earliest animals is shown in figure 41. As animal cells differentiated MTOCs, they lost their mitotic option. How many undulipodiated cells must have died, unable to return to their former state? How many times, in the history of our lineage, were cells incapable of restoring replicating MTOCs and thus predictable numbers of other organelles, chromosomes, and undulipodia per cell? Population regulation in the community comprising the protoeukaryotes must have been a persistent threat to existence.

These considerations lead us to an unavoidable hypothesis. Whatever the check-up process assuring the presence of undifferentiated MTOCs for mitosis and at least one copy of each complete organellar genome, that process simultaneously returned the community to a haploid state. The check-up process rendered fertilization obligate and, in our cellular lineage anyway, coupled sexuality with the reproduction of the multicellular individual.

The correlation between differentiation and meiotic sexuality is a reflection of the obligate requirement of the return to haploidy by meiosis (and the synaptonemal complex) as a check-up or organizational process. Mixis, the production of a single individual equally from two parents by way of the fertilization processes occurring at the level of fused cells or multicellular individuals, becomes a consequence of the need to preserve differentiation and development. Mixis itself is dispensable, and indeed many organisms have given up mixis while maintaining meiosis. There are many other ways of generating genetic variability and performing whatever other duties mixis itself performs. Mixis was never selected for directly. An inordinate amount of data has been collected in attempts to prove the selective advantage of mixis, especially in animals living in unstable environments (Bell, 1982). No such conclusion is available from the evidence: neither in constant nor in varying environments can mixis be shown to confer selective advantage over amictic (nonsexual) life cycles. Animals and plants that show no mixis, which is to say organisms that have lost the capacity to form offspring from different parents in sexual unions (such as apomictic or self-fertilizing plants), nevertheless

**Fig. 41.** *Trichoplax adhaerens,* the least morphologically complex animal still extant. (Drawing by Laszlo Meszoly.)

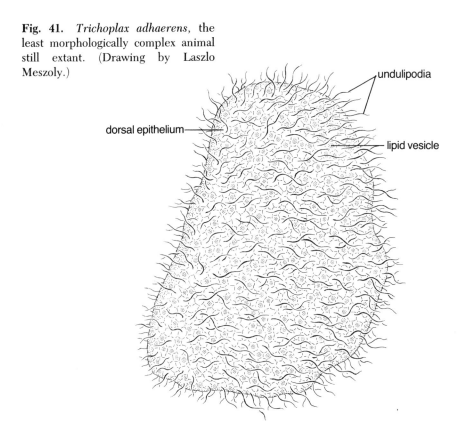

undulipodia

dorsal epithelium

lipid vesicle

retain meiosis. We suggest they do this to maintain differentiation. Outcrossing has never been shown to confer a definitive advantage on organisms. Leguminous plants are surely successful yet they generally are self-fertilized. Dandelions have bypassed mixis entirely, yet their "success" is undisputed. Perhaps the human social bias against inbreeding, as it exists in the incest taboo, has crept into and colored unstated scientific assumptions.

In any case, protists such as some hypermastigotes, autogamous *Paramecium,* and *Actinophrys,* engage in a seemingly meaningless meiosis. Just after the meiotic products are made, they fertilize each other. It is ridiculous to argue that the purpose of these meioses is outcrossing or mixis. If we accept that sexuality, in the sense of mixis, has been repeatedly lost we must wonder why meiosis has been retained. Cell, tissue, and organ differentiation are indispensable to organisms that develop from embryos: animals and plants. We hypothesize that the process of meiosis itself is intimately related to differentiation in animals and plants. We develop the idea that the meiotic prophase, with its special DNA and protein synthesis, is the inventory and reckoning, the insurance policy that checks and organizes the ancient microbial community of the cell preparatory to its reproduction into those careful populations described in traditional evolutionary biology as "individuals." Cloned and differentiated, the "microbes" become plants and animals.

It has already become clear that the synaptonemal complexes of meiotic prophase I are not necessarily related to mixis; indeed, they are not even necessarily part of meiosis (Stack and Brown, 1969). In the slime mold *Echinostelium minutum* synaptonemal complexes form but there is no alternation of haploid and diploid generations, no evidence for meiosis or fertilization, and no evidence for mixis (Haskins and Therrien, 1978). We believe these observations are entirely consistent with our concept that meiotic prophase, with its special DNA and protein syntheses, is required for differentiation in some protoctists and fungi and in virtually all animals and plants because it is a mechanism of intracellular genetic control. Synaptonemal complexes in the absence of meiosis in *Echinostelium* indicate that the genetic controls required for differentiation preceded, in evolution, the entire meiotic prophase I system indispensable to tissue-differentiated organisms.

# 12 · BIG EGGS AND SMALL SPERM

## Origin of Anisogamy and Gender

### THE UNSOLVED PROBLEM OF PROTOCTIST ANCESTORS
### OF ANIMALS AND PLANTS

The protoctist ancestors of the animals and plants never solved the problem, on the single-cell level, of how to retain both their motility and their ability to divide by mitosis. The contention that replicating and motile MTOCs are mutually exclusive would be falsified by the observation of a motile, undulipodiated animal or plant cell in the process of mitotic division. The cells that were motile by undulipodia and were forced therefore to relinquish mitosis were ancestral to animals and plants. They adopted multicellularity as a solution. Because a given single cell could not solve the problem of simultaneously retaining mitotic reproduction and undulipodial motility, it kept in contact with the mitotic cell from which it had originated. The swimming, dividing collective accomplished what was impossible for either cell alone.

Such two-celled microbes were antecedents of the protoctist species ancestral to animals and plants. Perhaps codescendants among living protists can be identified on the basis of the details of the pattern in the membrane around the mature undulipodia. Freeze-fracture analysis of this so-called ciliary necklace (Bardele, 1981) has generally shown a remarkably varied pattern in protists and constant pattern in animal cilia and their sperm (fig. 42). If protoctist cells can be located that display the ciliary necklace common to animal cells, we will have found a likely ancestor to the animal cell lineage. (As far as we know, no information on the ciliary necklace is yet available for plant sperm.)

Having become simultaneously motile and mitotic, our ancestors were

**Fig. 42.** Freeze-fracture micrographs of the undulipodial necklace. Protoctists (1. *Cyclidium;* 2. *Stephanopogon;* 3. *Opalina;* 4. *Monoceromonoides;* 5. *Joenia;* 6. *Trichomonas*) vary greatly in the pattern of the spherical protrusions of the protoplasmic fracture faces, whereas animals (7. *Ephydatia;* 8. *Aiptasia;* 9. *Typhloplana;* 10. *Schistosoma;* 11. *Aelosoma;* 12. *Ciona*) show uniformity, indicating both common ancestry and the conservative nature of undulipodial features. CP = ciliary plaques; N = necklace; FC = flagellar collar; FR = flagellar rosettes. Bar = 0.8 micrometers. (From Bardele, 1983; courtesy of Christian F. Bardele, University of Tübingen.)

trapped by their multicellularity. Variations on this problem were common in the protist world as single-celled protists evolved into multicellular protoctists. In several lineages the temptation to experiment with the numbers and interactions of the former microbes that comprised these collections of cells led to bizarre developments. These unusual life-styles are still present in today's parade of life in protoctists of all kinds: colonial vorticellids (ciliates with hypertrophy of undulipodia), colonial amoebomastigotes (acellular slime molds with a hypertrophy of nuclei and continual transformations from undulipodiated to dividing amoebal and multinucleate forms), colonial zoomastiginas (organisms that took up plastids, hypertrophied them, and became colonial euglenids), colonial diatoms (hypertrophied MTOCs?) displaying feats of motility (such as *Bacillaria* sp., which is structured like a set of stacked pillboxes and springs out like a series of jack-in-the-boxes; fig. 43).

In each protoctist lineage the presence of a variation on MTOC-based motility systems threatened the loss of mitotic division for any given cell. Any MTOC, once committed to form a centriole, undulipodium, or any other differentiated form of the organelle relinquished the availability of the MTOC for replicative (mitotic) use. Since nonmitotic cells, no matter how elegant their motility, would tend to die, selection was relentless on protoctists to retain the potential of mitosis. In the animal lineage, the problem was solved by the hitching up of mitotic and undulipodiated cells. By failure of separation of undulipodiated cells from those retaining mitosis, high heritability of the interacting former microbial community as protoctists was ensured. Weakness transformed into strength as new options opened up to microbial communities that become protoctists.

The protoctists that had combined mitosis and undulipodia through the ruse of multicellularity also evolved meiosis. From the beginning meiosis was associated with sex (mixis). For over 600 million years organisms in the animal and plant lineages have been paying a dear price for the privilege of differentiation. This price is meiosis. Sex, in the form of biparental mating and associated with meiosis since it arose, has been taken along for the ride. Although plants and occasionally animals can escape mixis, they cannot go for long without meiosis.

The onset of meiotic sex as a necessary component of differentiation had many consequences. Individuals aged. Machinations to search, locate, and fuse with fertilizable conspecifics evolved and were refined. Communities of former microbes, now eukaryotic cells, experimented in more and more extreme ways, forming new kinds of multicellular beings. In a flurry of recombinational activity that gave rise to a plethora of new pheomena, the new cells

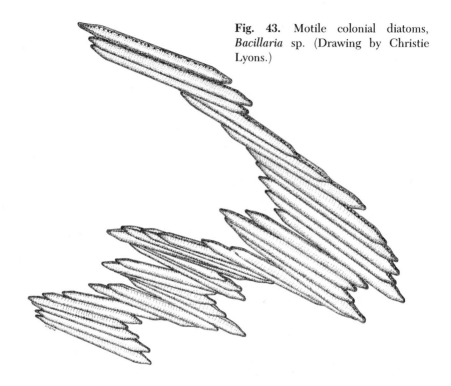

**Fig. 43.** Motile colonial diatoms, *Bacillaria* sp. (Drawing by Christie Lyons.)

deployed the genes, messenger RNAs, and other gene products, as well as the metabolites inside their various organelles, to form tissues and organs. Invertebrate animals with well-developed organ systems evolved. Algae with sexual life cycles, both rhodophytes and charophytes—multicellular green algae with a good fossil record—emerged. The genetic and protein synthetic systems of the component organelles were differentially used.

The results of organellar interactions on the level of genetic and protein synthetic systems are becoming familiar. Biochemists, geneticists, molecular biologists, and other investigators are revealing strange eukaryotic mechanisms of cell differentiation and control. The literature of eukaryotic molecular biology grows, but the practitioners of this science do not think of themselves as analysts of integrated symbioses. That they are studying a latter-day microbial community and its interactions has not yet been factored into their thinking.

How did the communities of symbiotic microbes become the orderly differentiated tissues and organs of animals and plants? The principle, suggested already by I. Wallin in 1927, involves differential growth and reproduction of the autopoietic systems comprising the eukaryotic cell. If our views are correct, at the basis of differentiation is a sort of "numbers game" in which internal cellular populations composed of nucleocytoplasmic, mito-chondrial, undulipodial, and sometimes plastid and other formerly free-living bacterial constituents reproduce at different rates. The quantitative fact of differential population reproduction leads to qualitative changes.

In some cases of differentiation, some genes themselves replicated at rates greater than the other genes in the cell. This sort of production of extra copies of genes is well known, for example, as the phenomenon of gene amplification in amphibian oocytes. In cells destined to become eggs the genes for the ribosomal components hypertrophy to the point of becoming visible micro-scopically. In other cases, one of the classes of microbes—now organelles—outgrew another. This is one way of looking at the far larger number of chloroplasts in leaf cells relative to the number in roots of flowering plants. Differential gene replication was followed by intracellular recombination in the remarkable phenomenon of cell differentiation. The "promiscuous DNA" of one class of organelles rampantly entered other organelles and replicated inside them (Lewin, 1984).

Extra copies of genes were easily produced by replication. Using the ancient ultraviolet-protective gene splicing and rejoining mechanisms, genes recombined inside the cell-as-community, making possible whole new com-binations of DNA, RNA, and protein. A given cell can produce thousands of genetic combinations. Each cell of the same original genetic constitution (derived from the same zygote) can produce a different set of these thousands of DNA, RNA, and protein molecules. (The ultraviolet-induced events of genetic recombination presaged the origin of the clonal immune system of vertebrates in which antibodies counteract an incredible number of proteins. The variable and constant components of the immunoglobulin genes, for example, IgG, are produced by way of gene recombination in the lymphocyte white blood cell lineage. See fig. 44.) We know that in this well-studied example of a differentiated cell lineage a stem cell—once, like a germ-line cell, containing one copy of each gene—in the end has a different genetic constitution. The new genetic constitution has differentiated by the hypertro-phy of the cells containing the appropriate variable regions of the IgG mole-cule in response to confrontation by the animal of external antigens. The differentiated cell, now producing specific IgG molecules against specific antigens, has abandoned forever its original genetic totipotency. To produce new animals capable of new variable responses to different antigens some process (we hypothesize meiosis) must ensure the retention of the intact genetic and developmental potential of the animal cell.

In some cases highly repeated classes of DNA led to differential protein

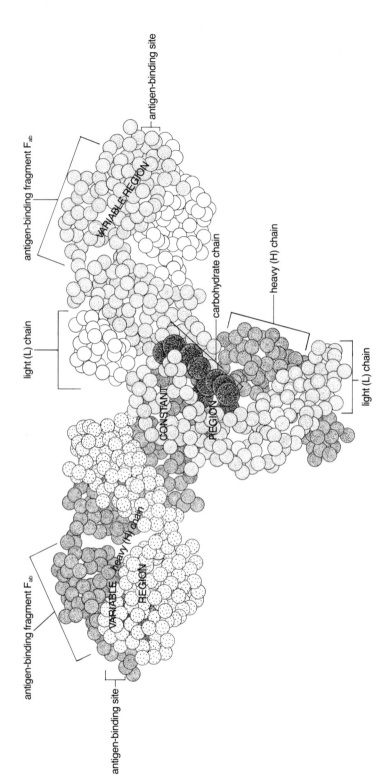

**Fig. 44.** Immunoglobulin G (IgG), a glycoprotein molecule. Each small circle represents an amino acid residue; large circles represent carbohydrates (such as mannose, glucosamine, n-acetyl glucosamine, and galactose). Because of different genes and differentiation involving DNA recombination generating the variable regions of these proteins, millions of chemically different immunoglobulins can be made in the serum of vertebrate animals. See Davis et al., 1980, for details. (Drawing by Christie Lyons.)

production. In other cases, the excess DNA inaugurated changes in cell architecture. Both of these phenomena were selected for in the protoctist lineages leading to the organized shapes of fungi, plants, and animals. Elegant and strange cell morphogeneses utilizing the replicating MTOCs were also part of the cell differentiation story. Cell morphogenesis, and therefore the development of asymmetries such as found in actinopods, nerve axons, and stellate melanocytes, were products of MTOC replication. In animals, division precedes differentiation. Cells divide and *then* grow undulipodia or axons. After they differentiate they never divide again. The rods, cones, hair cells, and myriads of other sensory structures derived from $9 + 0$ kinetosomes are also products of MTOC replication and differentiation (Atema, 1975).

Asymmetrical cell morphogenesis is often associated with movement and therefore best developed in protoctists and animals. Growing microtubules, associated with the kinetosomes of undulipodia (kinetids), became directly involved in the generation of cell form in many protoctists (such as the postciliary ribbons in the postciliodesmata subphylum of ciliates) and animal cells (such as the oyster gill ciliary rootlets and the microtubular campaniform, or bell-shaped, sensory structures of cockroaches). Kinetids, units of structure always involving at least one kinetosome as associated tubules and fibers, exemplify the relationship between numbers of organelles and cell differentiation. Most easily seen as units of structure on the ciliate cortex, they are present in all undulipodiated cells. They reproduce by development in at least some cases. Related protists differ primarily in the arrangement in number and position of kinetids (fig. 45). Kinetids are always composed of microtubules and therefore must be underlain somehow by MTOCs. Replicating genetic determinants in the form of excess copies of DNA or RNA must underlie the appearance of such supernumerary MTOCs.

At least some of the highly repeated DNA at the centromeric (kinetochoric) region of the chromosomes is not needed for protein synthesis but is directly involved in the meiotic process itself. *Stylonychia,* a heterotrichous ciliate, can live and grow by mitosis for an indefinite period of time with less than 90 percent of its DNA, but it cannot undergo meiosis without this DNA. The dispensable DNA is surrounded by new membranes, digested by nuclease enzymes, and lost in a complex process that occurs just after conjugation of these protists. This is the heterochromatic, nonprotein-synthesizing DNA of the differentiating, macromolecular, polytene chromosomes. It is likely that at least some of this DNA, dispensable for mitosis but not for meiosis, ensures the formation of the synaptonemal complex and its interaction with the MTOCs of the spindle (Raikov, 1982; Ammermann, 1973).

Much genetic imbalance and reshuffling apparently has been selected for

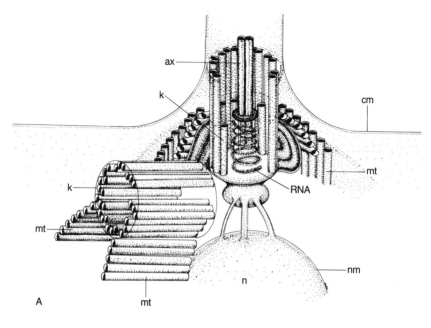

A

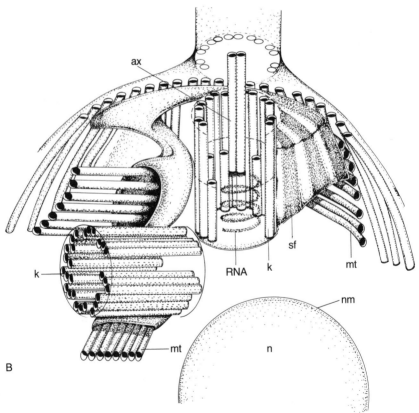

B

190

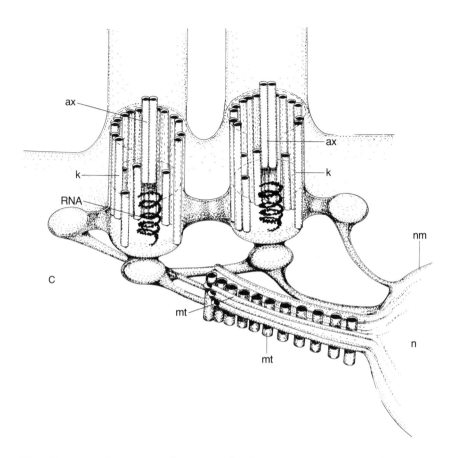

**Fig. 45.** Kinetid structure: three examples from protoctists. n = nucleus, nm = nuclear membrane, mt = microtubules, cm = cell membrane, RNA = ribonucleic acid helix of kinetosome, k = kinetosome, ax = axoneme, sf = striated fiber. A. Two orthogonally placed kinetosomes of the kinetid of the zoospore of a protostelid (a plasmodial slime mold) *Cavostelium bisporum.* B. *Ceratiomyxella tahitiensis,* another protostelid. C. The mastigote stage of the Cuban strain of the amoebomastigote *Paratetramitus jugosus.* For A. and B. see Spiegel, 1981. (Drawing by Steven Alexander.)

in autopoietic multicellular organisms. We hypothesize that the check-up system of meiosis must from time to time clean out all these excesses. Just as "garbage collection" is an indispensable part of the generation of complex computer programs, just as stock inventories must be taken in every large merchandising operation, meiosis, we feel, has to intervene in multicellular differentiated organisms to ensure the presence of at least one balanced, unshuffled, reproducible cell lineage. The presence of each component gene and organelle has to be verified. One of the early consequences of the obligate linkage between meiosis and differentiation was the independent origin, in so many lineages of multicellular organisms, of gamonts and different-sized gametes.

## GAMONTS

People are gamonts. So are the male pollen grains that form pollen tubes and the female stigma-style ovary that forms the egg of flowering plants. Conjugating *Paramecia* are gamonts also. The term *gamont* simply means an organism capable of entering a sexual encounter; one that can make haploid nuclei or, alternatively, haploid cells themselves capable of fertilization. The term is generally used for stages in the life cycles of protists such as foraminifera or apicomplexa. A gamont is an organism that, by mitosis or meiosis, produces cells or nuclei capable of fertilization. An *agamont* does not. It is an individual organism capable of reproduction, through mitosis of its component cells, to form another individual, but it is incapable of forming cells or nuclei that undergo meiosis or fertilization. Gamont stages alternating with agamont ones have evolved in many different lineages separately, underscoring again the general lack of correlation of sexuality and reproduction. The "agamonts" of plants such as mosses are the haploid gametophytes that grow and form dispersable spores. Many terms like *gametophyte, trichogyne, zygospore, macrogamete, sporozoite,* or *sporoblast,* specific to branches of biology, have been created to describe the various life cycle components— gametes and ramets (asexual generative organisms)—of protoctists, animals, plants, and fungi. The generalized life cycle for all eukaryotes can accommodate them all (fig. 46). All specialized terms can be translated into a consistent and unique set of descriptors (tables 12, 13).

## ISOGAMY

The first cells to fuse clumsily were unsuccessful cannibals, as we have seen in chapter 9. Very much like each other, from the same population of

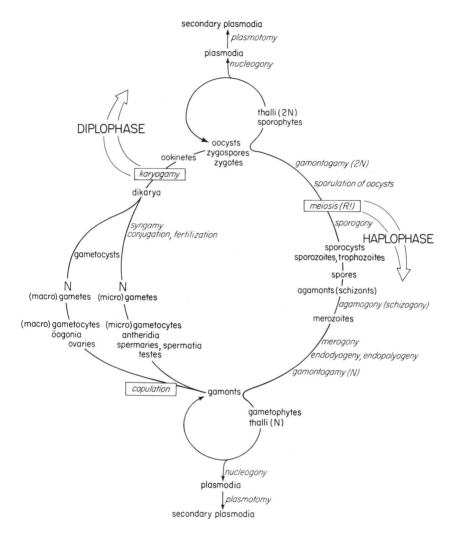

**Fig. 46.** The eukaryotic life cycle. The processes of reproduction and growth associated with stages in the life cycle are shown in *italic* type; the stages are shown in roman type. (Drawing by Laszlo Meszoly.)

**Table 12.** The Terminology of Sexual Processes

| PROCESS | DESCRIPTION |
|---|---|
| Reproduction | The process that augments the number of cells or organisms |
| Sex | Any process that unites genes (DNA) in an individual cell or organism from more than a single source |
| Recombination | Breakage and reunion of DNA molecules |
| Syngamy | Contact or fusion of gametes (cells) |
| Conjugation | Contact or fusion of gametes or gamonts (cells or organisms) |
| Gametogamy | Fusion of gametes (cells or nuclei) |
| Gamontogamy | Aggregation or union of gamonts (organisms); equivalent to mating, conjugation |
| Karyogamy | Fusion of gamete nuclei |
| Crossing over | Breakage and reunion of DNA of homologous (nonsister) chromatids in meiosis (or mitosis) |
| Amphimixis (mixis) | Syngamy or karyogamy leading to fertilization to form an individual with two different parents; equivalent to outcrossing, outbreeding |
| Amixis | Absence of meiosis and fertilization at all stages in the life cycle; equivalent to asexual reproduction |
| Apomixis | Altered meiosis or fertilization such that amphimixis is bypassed (e.g., parthenogenesis) |
| Automixis | Syngamy or karyogamy of nuclei or cells deriving from the same parent; equivalent to selfing, extreme inbreeding, autogamy |
| Parthenogenesis | Development of eggs or macrogametes in the absence of amphimixis |
| Arrhenotoky | Parthenogenesis producing haploid males and amphimictically produced diploid females; an incomplete fertilization event, in which unfertilized eggs become males |
| Heterogony | Parthenogenesis in the absence of karyogamy stimulated by sperm of a second species |
| Thelytoky | Parthenogenesis in which diploid individuals are formed by karyogamy of the egg with its own female pronucleus |
| Tychoparthenogenesis. | Occasional parthenogenesis |
| Gametic meiosis | Meiosis immediately preceding gametogenesis (characteristic of members of Animalia kingdom) |
| Zygotic meiosis | Meiosis immediately following zygote formation (characteristic of members of Fungi kingdom) |
| Haplodiplomeiosis | Meiosis preceding an extensive haploid life cycle in organisms that also have extensive diploid life-cycle phases (characteristic of members of Plantae kingdom) |

protists living in the same habitat at the same time, these reluctant mates could be considered isogamous: equal haploids entering a "fertilizing" event. The trends from isogamy (the union of haploid cells or nuclei that are equal in appearance) to anisogamy (the union of haploid cells or nuclei that are different in appearance), occurred in protoctists, plants, and fungi many separate times. The smaller of a pair of gametes is dubbed "male." Male sex cells are so designated, however, not only because of their smaller size but because of their propensity to move on their own.

Trends toward anisogamy were notable in most, if not all, of the major groups of meiotic protoctists. Male haploids, or sperm, became smaller, more motile, and more numerous. The female gametes, or eggs, became increasingly large and stationary. Genera of green algae, ciliates, apicomplexa, and even rhodophytes (red algae, in which the smaller "male" cells are not motile by themselves) have isogamous and anisogamous member species.

Indeed, the practical definition of *male* is simply an organism (gamont) that produces moving, relatively small gametes, often in profusion, but the word may also be used for one of those gametes (or microgametes) itself. The term *female*, by general definition, usually refers to those gametes which stand still or those gamonts which produce sedentary gametes. Relieved of the need to seek out cells with which to fuse, gametes or gamonts in many species were specialized to devote their time to making fewer products, thereby retaining more nutrient and expending less energy in motility.

Isogamy, the fusion of two equal-sized haploid cells or nuclei in the formation of the zygote, is found today only in certain protoctists and fungi. Even in groups of chlorophytes, ciliates, diatoms, and ascomycotes the trend from isogamy toward anisogamy is apparent in the merger of haploids. The morphologically simpler organisms in any related lineage tend toward isogamy and the more complex toward anisogamy. It was concluded many years ago (for example, Jennings, 1920) that isogamy is primitive relative to anisogamy. As long as the equality of the parental nuclear contributions is maintained, division of labor leads to efficiency: one parent cell stores food and stays put and the other loses all its excess baggage and moves around. Even though, by definition, the entity that travels is male and the sedentary partner is female, sexual differentiation is not that easily characterized. For example, the female part in ascomycotes, a thread of hypha called the "trichogyne" (Greek, "female thread"), actually travels to the male. It grows out, contacting the male hypha. After fusion, the nuclei from the male and female partner return to the female and a new sedentary structure, the ascus, on the female, is developed.

In the evolution of anisogamy, all excess weight can be lost from the male to lighten his burden. But he must retain his *sine qua non:* his multigenomic contribution to the community of which he becomes a part. One copy of each of the genes, including those of the organelles (if the female does not provide them), and the minimal set of materials for the expression of those genes must be maintained. Figure 47 shows the tendency toward anisogamy in several different lineages.

**Table 13.** The Terminology of Sexual Structures

| STRUCTURE | DESCRIPTION |
|---|---|
| *Prokaryotes* | |
| Genophore | DNA, genes of the bacterial genome (observed as light fibrous region or nucleoid in bacteria) |
| Chromonema | DNA, not complexed with protein, that comprises the genophore |
| *Eukaryotes* | |
| Chromatin | Histone-complexed DNA that comprises chromosomes |
| Chromosome | Chromatin gene-bearing structure of eukaryotes; at least two per cell |
| Gamete | Haploid cell or nucleus requiring fertilization for further development |
| Oogamete | Large gamete, macrogamete, or egg; if motile called an "ookinete" |
| Microgamete | Small gamete, or sperm |
| Zygote | Cell or nucleus that is the product of syngamy or karyogamy |
| Kinetochore | Spindle fiber attachment or centromere, that portion of the chromosome that connects two chromatids and contacts the spindle or the nuclear membrane |
| Synaptonemal complex | Protein matrix that binds homologous chromosomes |
| Dikaryon | Binucleate cell (if it is known that the two nuclei are from different parents, the dikaryon is a *heterokaryon*) |
| Gamont | Organism or cell capable of forming gametes |
| Agamont | Reproducing organism or cell incapable of forming gametes |
| *Both Prokaryotes and Eukaryotes* | |
| Genome | The total genetic material of an organism |
| Organelle | Morphologically distinguishable substructure of a cell |

## GENDER

As long as meiosis-fertilization cycles have been maintained, mechanisms to ensure contact between haploids have also been maintained. The differentiation that distinguishes the two haploids that fuse has traditionally been called "sexual dimorphism." The term has been used to mean any visible differences between gamont partners. Since nuclei, whole cells, and multicellular individual organisms may fuse, sexual dimorphism may be at any of these levels. The term *mate* is used to indicate the gamonts, the sexual partners, the nuclei, cells, or individuals that enter the fusion. Mates are distinguished by mating types. The basic rule of gender or mating type is definitional: organisms do not mate successfully and produce living offspring with others of the same mating type.

In species of animals, plants, and others that require meiosis and fusion of haploids at the time of reproduction of the individual, speciation itself is defined by sexual compatibility. Organisms lacking sexual compatibility, the capacity to make new viable organisms through sexual fusion, are considered to belong to separate species. Each of these millions of species has its specific signals of mate recognition. The entire complex of signals of mate recognition, for any species of organism, is defined as gender. It is obvious then that each time a new species of animal or plant appeared a new change in gender also evolved. Determination of gender (or sex determination, as it is called in the literature) is, then, incredibly polyphyletic: new means or variations on the theme of distinguishing and recognizing mates evolved at least as many times as new species evolved. In those species in which gamonts as well as nuclei and cells fuse, specific recognition between mates occurs at three levels, those of the nucleus, the cell, and the individual multicellular organism. A huge number of mechanisms, many still unstudied, must exist for specific gender recognition. If the problem of the origin of sex is posed as the problem of the origin of gender, it becomes not a single problem at all, but millions of problems. Whatever the ultimate molecular basis of the mechanisms, explanations of gender must include the recognition of mates, fusion of gamont bodies (gamontogamy), fusion of haploid cells (syngamy) and ultimately of their nuclei (karyogamy), and, in species that cross over, the splicing and rejoining of their DNA (genetic recombination) as well.

## MINIMUM REQUIREMENTS FOR GENDER

The minimal requirements for differentiating mating types are indeed minimal: they can be the presence of a protein on the surface of the undu-

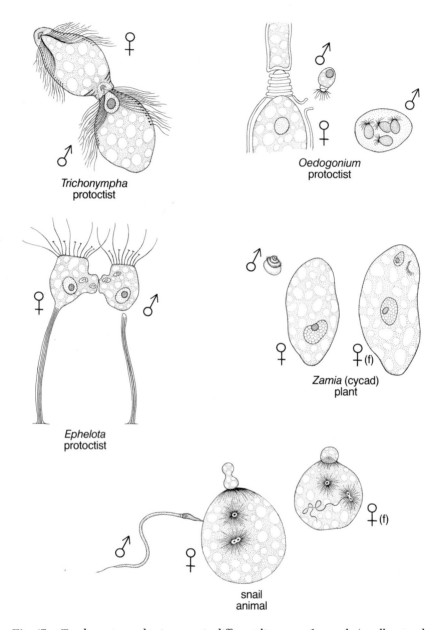

**Fig. 47.** Tendency toward anisogamy in different lineages. ♂ = male (smaller, tends to travel); ♀ = female (larger, tends towards sedentary behavior); ♀ (f) = female after fertilization. (Drawing by Laszlo Meszoly.)

lipodia (as in ciliates) or the existence in one position but not in another of a transposable piece of DNA (as in yeast). At the minimum they involve only the difference of expression of a single gene at a single locus.

As protoctists, plants, fungi, and animals became increasingly long-lived and multicellular, the tendency toward sexual dimorphism became increasingly marked. Gender distinctions, from the mating type "plus" in *Chlamydomonas*, to the booming prairie chicken, the bull seal, and the malodorous female ginkgo tree, evolved millions of times. Phenotypically distinct sexual types may result from minute genetic differences. For example, the allele of a gene determines two opposite mating types (I and II) in the group A stock of *Paramecium aurelia* (Beale, 1954).

Many single eukaryotic cells or multicellular organisms whose component cells divide by mitosis can produce two or more kinds of gametes or gamonts. That is, a single individual, by mitotic division and growth, can produce offspring of more than a single mating type or gender. These offspring (gamonts) or their representative cells (gametes) later are capable of fusion with each other. This is the phenomenon of hermaphroditism in animals, monoecism in plants, and heterothallism in fungi. One parental cell forms offspring of more than a single mating type in many organisms: a single fungal spore often gives rise to its own offspring, which then mate with each other without any sexual fusion. One of the best studied cases of sex determination is in the ascomycote *Saccharomyces cerevisiae* (baker's yeast).

This kind of sex determination involves only the movement of a small piece of DNA from one place in the genome to another. The gene products of yeasts of opposite mating type ($\alpha$ mates with $a$) are oligopeptides only 11–13 aminoacyl residues long but made from a precursor of 165 amino acid residues. Each $a$ cell (but not $\alpha$ cells or $a/\alpha$ diploid cells) on its surface has a receptor protein of molecular weight about 100,000 for the $\alpha$ oligopeptide. The $\alpha$ peptide stops growth of $a$ cells or prevents them from entering DNA synthesis stages. It then induces in $a$ cells thin walls with fuzzy extracellular coats probably by changing intracellular concentrations of cyclic ATP. The glucan layers of the mating cells become progressively thinner as the walls are energetically digested by enzymes and (when the same sorts of changes are induced in the partner) the cytoplasms fuse (Saier and Jacobsen, 1984). Beginning with such minuscule molecular differences between gender there have evolved, with time, many lineages of organisms that formed single individuals with large numbers of behavioral, physiological, and other differences. Most of these individuals can make only one kind of haploid nucleus—either male or female.

We are, of course, accustomed to the ubiquitous terms *female* for one mating type and *male* for the other. Examples from all kingdoms reveal, however, that the well-known heterochromosomal mammalian XX-female and XY-male gender determination mechanism lies at one end of a spectrum—the more rigidly determined end. Even these mating types of man and other vertebrates, as we shall see below, are not immutable. Mating types of nearly all other organisms tend to be much less fixed than those of vertebrates. Nor do there have to be only the two genders "male" and "female."

There are, of course, many examples of chemically and environmentally influenced sex determination even in the zoological literature (Ghiselin, 1974; Hapgood, 1979). Perhaps the most remarkable and well-studied example of labile gender is described in the literature on ciliates. It is probably the only example known in which the genetics, biochemistry, morphology, and behavior are all understood even in principle. Populations of the ciliate *Paramecium multimicronucleatum*, like many paramecia, display several different mating types. Mating types can change or be lost according to the time of day (Barnett, 1966). During the day one clone will be of mating type 1 and will mate only with mating type 2. At dusk members of the clone will not mate at all. By nightfall the mating type 1s will have changed into 2s and thus will refuse to mate with other mating types 2s. By changing the lighting conditions, making "dawn" and "dusk" come at different times, the formation of different sexes or mating types in these paramecia can be controlled. Mating type is determined by the presence of proteins on the surface of the cilia. These proteins are such strong attractants that the cilia from organisms of compatible mating types will "mate" all by themselves. That is, even when sheared from the rest of the protist, the cilia will stick to those of the "opposite sex" (that is, complementary mating type), as if the rest of the protist cell were still there (Watanabe, 1977).

## CHROMOSOMES AND GENDER

There are many steps in the development of gender in animals and plants. Although the chromosomal mechanisms determine the gender of vertebrates, the gender displayed by an individual depends on the chromosome's DNA being read by messenger RNA to produce proteins. These proteins produce differentiated cells that in turn produce small molecules, such as steroid hormones, that travel to other cells. The steroid hormones enter their target cells, bind with chromosomes there, and permit the differential production of gene products, messenger RNA, and protein (Witzmann, 1981). Interruption or, in some species, environmental influences on these processes at any level can lead to severe alterations in the gender-generating process.

We shall mention only a few examples of the relation between gender determination and gene expression. In the usual case, humans with both an X and a Y chromosome in each of their cells are men. However, passably ordinary looking women have been observed with this chromosome constitution. These women are sterile; they are and remain genetically male. They have suffered a mutation on their X chromosome that has led to the production of female steroid hormones and the development of testicular feminization. From figure 48 it can be seen that at crucial developmental stages, about eight weeks after fertilization, massive quantities of hormones determine the masculine or feminine developmental pathways.

A single individual, independent of the genes for one or another sex carried in the nuclei of its cells, can be either male or female. This is due to the influences of hormones on development. The effect of hormones depends on environmental or developmental cues. For example, *Ophryotrocha*, when young, is a sperm-producing male. As it grows in length it becomes an egg-producing female. This marine annelid worm can be cut in two and survive. If an older female is cut it develops again into a male. Thus length seems to be more important than age in determining the sex of this animal. Although chromosomes entirely determine the genetically transmitted gender in vertebrate animals, hormones determine the actual functioning gender. Male eggs of Japanese killifish, if treated with estrogen, develop into functioning and even fertile females. Such females can be mated with normal males (XY crossing with XY) and produce live offspring (XX females, 2XY males, and YY males). These YY males survive. If treated with hormones they become females. That they are really YY individuals genetically can be verified by crossing them with YY males and showing that all offspring are YY males. Such observations emphasize the fundamental lack of rigidity in gender determination and the extraordinary similarity between males and females, even in vertebrate animals.

The differences between gamete nuclei, cells, or gamonts that distinguish organisms as mates can, however, be enormous. The story of evolutionary biology has very much been the story of increasing complexity of gender determination. The annals of biology are replete with details, from the transposed DNA that determines the *a* or α mating type in yeast to the distinguishing features between men and women. In summary, then, we may say that when meiotic sex first evolved, isogamy was the rule. Sexual partners were nearly identical in appearance, as many sexual protoctists still are today. Eventually a host of different genders and their determining mechanisms evolved.

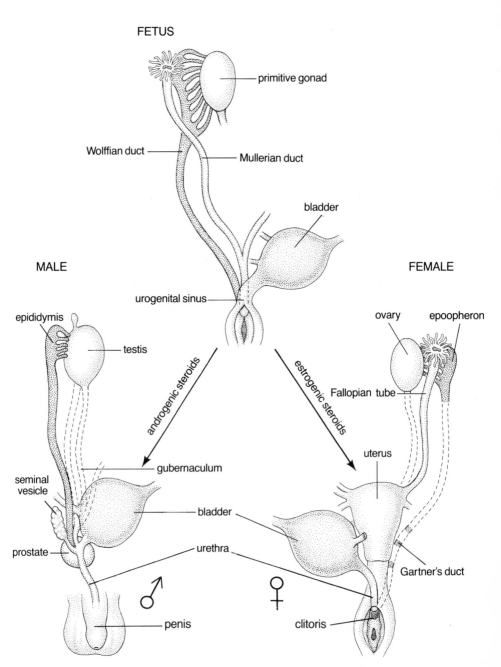

**Fig. 48.** Differentiation of human sex organs. This diagram shows the influence of steroid hormones on differentiation of human male and female sex organs from their developmental antecedents. Dotted lines indicate absence of tissue in stage shown. Testes tucked in scrotum in adult male. (Drawing by Laszlo Meszoly.)

In Darwin's epic of 1859, *The Origin of Species,* he discusses the many themes of the mechanisms of sexual differentiation and the behavior of the gender partnerships in the production of offspring ("sexual selection"). These themes have been explored extensively in the evolutionary literature (Bell, 1982; Hamilton, 1972; Jennings, 1920; Hapgood, 1979; Ghiselin, 1974; Williams, 1966). Influential attempts have been made to mathematize both theories and observations (Maynard Smith, 1978). We have little to add except clarification. The fact that these themes—genetic recombination, meiotic sex, gender determination, and reproduction—have been inextricably linked in vertebrate animals, coupled with the fact that *we* are vertebrate animals, has led to great confusion. In our opinion, the confusion has generated the statement of pseudoproblems in the scientific literature. We believe that the origin of sex is a large and varied set of different problems. It is not easily amenable to unitary mathematical treatment. We assert that gender is an epiphenomenon of meiotic sex and cell differentiation. Gender is determined in many different ways with different degrees of rigidity. Genders have changed many times by mutation and gene expression. Gender-changing mutations occurred as heritable alterations in the genes of the original community of microbes that comprised the eukaryotic cells. Changes in the genetic basis of gender determination occurred as many times as speciation itself in the rise of obligately sexual species. In our final chapter we explore the consequences of our views for the statement of scientific problems concerned with the origin and maintenance of sex.

# 13 · THREE BILLION YEARS OF SEX

## Common Evolutionary Legacy and Strange Variations

### A POINT OF VIEW

As was astutely realized by Langridge, the complex problem of the origin of sex is ultimately connected to the problem of pre-Phanerozoic genome evolution (Langridge, 1982). We now summarize the historical perspective developed in this book that we share with Langridge in the hope of broadening the base for further discussion. In this final chapter we unite our conclusions with his implications; this should provide the basis for further explanation and research. We apply our general conclusions to some specific examples: trends of sexuality in symbiotic associations and the use of our analysis for understanding ciliate differentiation and loss of sex in ciliates and other formerly sexual groups.

### AUTOPOIESIS AND REPRODUCTION: DIFFERENTIATION AND SEX

One can imagine autopoiesis without reproduction: it would be accomplished by nutrient intake and nucleic acid and protein synthesis in a persistent, self-maintaining individual cell. Why, then, after the evolution of autopoiesis did the first bacterial cells reproduce by division? DNA tends to replicate, produce more protein, and replicate again. Perhaps in the earliest protocells surface area per volume became too small to support the newly replicated DNA and its nutrient needs. The requirement for cell reproduction may be related to an early sort of error correction. During the replication of complementary DNA strands there is a kind of confirmation: the polymerases by nature of the replication process determine that the 3' strand is still intact and complementary to the 5' strand (Kornberg, 1980). Replication of nucleic acid molecules leading to reproduction of cells may have been a consequence of the necessary error correction of complex informational systems. Likewise, as we have seen in chapters 10–12, meiosis may have evolved as a further error-correcting device for differentiation in the microbial community we call a cell. We may never know the best explanation for the origin of reproduction and sexuality. We can only state hypotheses and explore the consequences.

Let us review the assumptions we have made and the history we have developed in this book to see how it differs from conventional wisdom regarding the origin(s) of sex.

First of all we have assumed that autopoiesis is a prerequisite for reproduction and that the minimal autopoietic unit is a cell. Reproduction, in turn, is seen as a prerequisite for sex. Reproduction preceded all kinds of sex and is not an intrinsic part of the sexual process. Since sex, in the form of mixis, is not selected for directly, the question, "Why, if asexual beings can have far more offspring than sexual ones, are there so many more sexual animals?" is not a valid scientific problem. Many complex organisms do not have any asexual options at all.

Meiotic sexuality was never selected for because it generated more variation than asexuality did. Meiosis evolved as a cyclical relief of diploidy and was maintained in many species first because of seasonal or other alternating environmental conditions and later due to its obligate association with the development of differentiation. The lack of inherited variation was not a limiting factor in the evolution of new life forms. Inherited variation, achieved by many means, is already in adequate supply. It is generated by mutation as well as plasmid-, genophore-, and viral-mediated recombination in prokaryotic populations. In early eukaryotes variation was generated also by modular processes such as symbiont acquisition. Symbiosis has always been a major mechanism for generating important new inherited variation (Wallin, 1927; Pyrozinski and Malloch, 1975; Margulis, 1981; Schenk and Schwemmler, 1983). Further variation was then produced by the interaction of the components of the genetic systems of the various symbionts comprising the cell. After the evolution of chromosomes, variation was generated by changes in the number and morphology of chromosomes (for examples in plants, see Jackson and Hauber, 1983) or by karyotypic fissioning (Todd, 1970). In many populations variation is generated by an enormous repertoire of gene expression of members of a clone. In other populations variation is minimized by failure of the variants to survive. It has not even been proved that mictic populations harbor more genetic variation than amictic ones (Bell, 1982). That is, populations with a great deal of measurable variation may be amictic—"nonsexual."

Sexual selection in animals was enormously elaborated because of its association with reproduction, which is obligate. The evolution of genders ensured reproduction in differentiated organisms (animals and plants). Gender determination minimally involves an environmental effect on gene expression or the rearrangement of a single small piece of DNA. Chemically, sex differences may be attributable to small peptides. At the maximum, gender determination is far more extensive. It involves processes at the levels of the gene, chromosome, cell, tissue, organ, organ system, and behavior of an organism.

Any explanation of the origin, evolution, and maintenance of gender differences must concern itself with the populations of organisms, the times and places in which these gender differences evolved. For this reason the problem of the origin, evolution, and maintenance of sex is a set of multiple and complex biological problems. Most data concerning these problems are not yet adequate for mathematical treatment.

The original sexual events, at the DNA recombinational level, accompanied the origin of bacterial DNA repair systems. The DNA-processing enzyme systems involved in cell differentiation were retained and reused in the origin of chromatin and in crossing over. Using DNA-processing enzymes, genes have been transferred from organelle to organelle, primarily the movement of genes from former symbionts—mitochondria, plastids, MTOCs—to the nucleus (Gray, 1983; Thornley and Harington, 1981; Obar and Green, 1985; Lewin, 1984). These enzymes also have been used extensively in gene amplification (repeated expression of the same gene) and degradation processes associated with differentiated tissues in animals (namely, the immunoglobulin system). Prokaryotic sexuality was a preadaptation for tissue differentiation; cannibalism followed by indigestion (inability to digest conspecifics) was a preadaptation for meiotic sex.

## SEX AND SYMBIOSIS

The concept that sexuality is maintained by natural selection is so deeply ingrained in modern evolutionary thought that it generally goes unquestioned as a premise. If the hypotheses presented in this book are correct, not only is the question of the maintenance of sex as stated a nonscientific one, but the very construction of the problem in this way leads to intellectual mischief and confusion. (Comparable to the now obsolete scientific problem of which blue-green algae are ancestral to plants, the question of what forces of natural selection maintain sex is intrinsically obfuscating.) Any data relevant to the issue of the distribution of variation and sexuality, of course, warrant explanation in alternative terms.

For example, in a brilliant and fascinating paper that reviews a large quantity of literature on symbiosis, parasitism, species diversity, and sexuality, Law and Lewis (1983) claim:

> A critical problem in evolutionary biology is to identify the forces of natural selection that maintain sex in populations of living organisms. (Here we use sex in a very broad sense to include all those processes involved in reshuffling chromosomal genomes in both pro- and eukaryotes.) That such forces do exist is self evident, in view of the widespread occurrence of sex in nature. However, it is far from clear what they may be (Maynard Smith, 1978). (p. 250)

These authors have derived some generalizations from their collection of an immense amount of data relevant to the question of the sexuality of partners in symbioses. They distinguish between exhabitants (outside members, facing the environment) and inhabitants (endosymbionts, partners living inside the exhabitants). Taken together, their data indicate that sexuality tends to be maintained in the exhabitants (for example, corals and marine invertebrates) but lost by the inhabitants in mutualistic associations. Furthermore, inhabitants tend to show very little species diversity, whereas species diversity is truly impressive in the exhabitants (Law and Lewis, 1983, table 2, p. 255). Law and Lewis show that variation and sexuality are conspicuous in the external but not in the internal partners in mutualistic associations. On the other hand, their data also indicate that sexuality tends to be retained in the inhabitants of antagonistic associations (parasitic protists and worms, and pathogenic fungi, for example). After recognizing that the capacity to generate new resistant recombinants is crucial in parasitic inhabitants that must resist host attack, Law and Lewis explain their results by claiming that endosymbionts are not subject to variable environments and thus sexuality is not maintained in them. Exhabitants and antagonized inhabitants, forced to face the rigors of changing environments, must maintain sexuality as their mechanism for the generation of genetic variation.

Yet these authors, experts on fungus-plant associations, have raised several criticisms to their own conclusions. Ectomycorrhizae, associations between basidiomycote fungi and woody plants, are enormously diverse. In these cases both types of partners, primarily forest trees and mushrooms, retain meiotic sexuality and extensive differentiation. To explain the failure of those basidiomycotes in association with hosts to lose sex, relative to basidiomycotes unassociated with hosts, Law and Lewis note, "The intimacy between the partners is much reduced, relaxing selection pressures for constancy in the 'inhabitants'; concurrently most of the contacts of the 'inhabitants' are external to their hosts so they are experiencing the full force of selection from their antagonistic biotic environment" (p. 263). That is, ectomycorrhizae are more "exhabitants" than they are "inhabitants," therefore they do not conform to the rules developed for inhabitants.

In addition, these authors wrestle with another observation: many truly "inhabitant" symbionts have stages external to their hosts in which they must survive in widely varying environments. Such symbionts should therefore display sex—but they do not. Much of the argument Law and Lewis present is based on examples of asexual organisms such as *Rhizobium*, a bacterial symbiont of the roots of sexual and diverse leguminous plants. The argument depends also on association between plants of the heath, members of the family Ericaceae, informally called "ericoids," and their vesicular-arbuscular mycorrhizae. (These fungal inhabitants are endosymbionts—primarily of the roots—of more than a quarter of a million diverse and predominantly sexual host plants.) The rhizobial and endomycorrhizal inhabitants are asexual in the sense that they show no evidence of gender differentiation and mixis. Yet they must in all of these cases face, at some point in their life cycles, hostile external environments. That is, these inhabitants are exposed to extremely variable outside environments and therefore, according to Law and Lewis, should display sexuality. Law and Lewis, then, need caveats for their argument, or, as we suspect, the argument itself is flawed because it is based on the commonly asserted and intuitively appealing but ultimately unjustifiable assumption that hostile environments maintain sexuality.

Some of the most variable organisms facing continuously changing hostile environments are entirely asexual. Among the most dramatic parasites in their rapidity of alteration are members of the group Kinetoplastida (*Trypanosoma, Leishmania*). Kinetoplastids vary enormously, and they brilliantly evade the antibody responses of their hosts by generating inherited variation in the form of resistant proteins on their surfaces. They do this without any sexual life cycle stages at all. Another striking counterexample to the supposed tendency toward asexuality of inhabitants in a constant, intraorganismic environment is that of mitochondria, which have been known to undergo prokaryotic-style sex inside mitotically dividing (asexual) yeast cells (Gillham, 1978).

It seems clear to us that it is meiosis, not necessarily mixis, that is linked to cell, tissue, and organ differentiation. Sexuality in our view is retained by antagonistic parasites because they require extensive tissue differentiation to complete their life cycles. In the types of parasites, primarily of plants and fungi, cited by Law and Lewis, mutation, recombination, temporal differentiation of distinct morphologies, and chemistries probably all play roles in resistance to host attack. Indeed, single gene mutations, not sexuality per se, have been amply demonstrated to confer both pathogenicity and resistance in both host and parasite in plant-fungus relationships (P. R. Day, 1974; the "gene-for-gene" hypothesis). As we have repeatedly seen, meiosis—not mixis—is retained because of its relation to differentiation. Law and Lewis are noting further examples of this. In short, Law and Lewis simply have not been able to defend their ideas. There is no evidence that biparental sex is maintained in organisms exposed to highly unpredictable external environments and lost in symbionts exposed to highly predictable environments.

## CILIATE DIFFERENTIATION

Members of the phylum Ciliophora, a relatively well-known group of protoctists, are primarily single cells. Because some of the species, such as *Tetrahymena, Paramecium, Stylonychia, Oxytricha*, and *Colpoda*, are extremely easy to cultivate in the laboratory, an enormous literature has discussed the growth, development, heredity, and behavior of these organisms (Corliss, 1979; Giese, 1973). The genus for which perhaps there is the most detailed information at all levels is *Paramecium* (Beale, 1954; Kung et al., 1975; Saier and Jacobson, 1984). Ciliate cells are highly asymmetrical and structurally complex. As sexual organisms they conjugate, exchanging haploid nuclei in such a way that reproduction (transverse cell division) and sexuality (conjugation) are entirely separated in time and therefore can be studied as separate phenomena.

Ciliates are defined as eukaryotic cells with dimorphic nuclei (macro- and micronuclei) and a cortex, the fairly rigid surface layer about one micrometer thick. The cortex is composed of kinetids. Kinetids are best known in ciliates, but they are characteristic of all undulipodiated cells. Only in certain ciliates is there a fairly complete idea of how kinetids reproduce (fig. 49). Kinetids in ciliates generally include the 9 + 0 kinetosome, the associated axoneme, the kinetodesmal (or striated root) fibers, the membranous structures (parasomal sacs), the transverse and longitudinal fibers, and sometimes other structures surrounding and oriented toward the kinetosomes. All these structures of course must be produced to form the offspring kinetid. In sperm, hypermastigotes, and chytrid zoospores, for example, kinetids are quite different and tend to be species- or genus-specific; their reproduction is less well

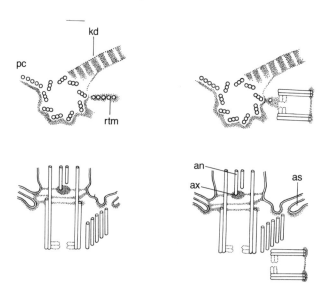

**Fig. 49.** Reproduction of kinetids. kd = kinetodesmal fiber; rtm = radial transverse microtubules; pc = postciliary microtubules; as = alveolar sac; ax = axosome; an = axoneme. At left, parent kinetid only; after development, at right, both parent and mature offspring kinetids can be seen with kinetosomes parallel to each other. (Drawing by Laszlo Meszoly.)

understood. All ciliates—at some time in their life cycle—have well-defined kinetids with one, two, or three or more kinetosomes per kinetid.

The morphology of the kinetid as a unit of structure is proving to be a remarkably stable and consistent criterion for relatedness in ciliates. Ciliate speciation is best understood as a change in the ultrastructural morphology of the repeat unit itself (the kinetid), a change in the distribution of kinetids (fig. 50). Kinetids, in all their patterned detail, reproduce, but not by direct division. They develop offspring kinetids often in proximity to parent ones; in many ciliates new kinetosomes are at 90° angles from old ones (fig. 49).

In ciliates the large macronucleus is the physiological nucleus of the cell that hypertrophies DNA, produces RNA, and directs protein synthesis. The ribosomal gene products—ribosomal subunit proteins and RNAs—are synthesized under macronuclear DNA direction. At least one kind of macronuclear RNA, the 26S ribosomal RNA that is essential to the large subunit of the ribosome, is capable of catalyzing its own intervening sequence (Cech, 1983; Abelson, 1982). (*S* in 26S stands for *Svedberg*, a unit of size based on how fast a particle sediments in a density gradient.) In this way the ribosomal RNA can prepare itself for assembly into mature ribosomes in the absence of

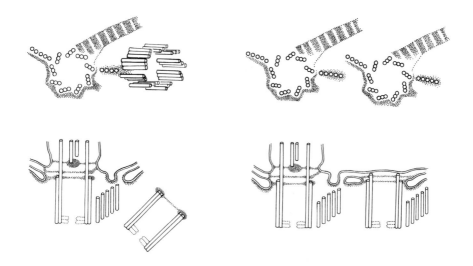

enzyme proteins. Enormous quantities of gene amplification are involved in macronuclear differentiation; sometimes several thousand copies of certain genes are present. The micronuclei retain single copies of each gene. Only the micronuclei can undergo meiosis and only the micronuclei can produce, by a sequence of steps in differentiation, macronuclei.

In some ciliates (the karyorelictids), the macronuclei cannot divide after differentiation. In each generation new micronuclei, produced by division from parental micronuclei, must differentiate macronuclei from micronuclei (Raikov, 1972). Thus, early in the history of the ciliate lineage a problem (analogous to the problem of the retention of motility and mitosis in animal ancestors) faced ciliates. How could differential gene replication and degradation be used for differentiation (in the production of the macronucleus) without the loss of the ability of the macronucleus to divide being lost? This problem, how to differentiate and also divide, was not solved by the earliest karyorelictids, but it was solved both by later karyorelictids and independently in other lineages (Lynn and Small, 1981). The solution of macronuclear division did not involve standard mitosis with standard chromosomes but an MTOC-related microtubular process of segregation of hypertrophied chromatin (Raikov, 1982).

Because ciliates, unlike most other eukaryotes, have so many kinetids, they display extremely regular and observable patterns on their cortices.

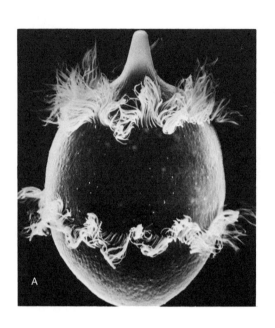

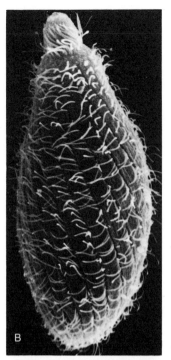

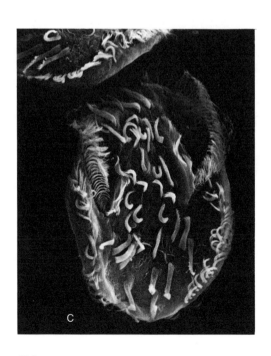

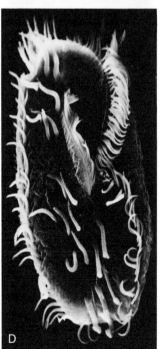

**Fig. 50.** Ciliate speciation. Scanning electron micrographs. A. *Didinium.* (Courtesy of E. Small.) B. *Tetrahymena.* (Courtesy of E. Small.) C. *Stylonychia mytilus* cortical doublet. (Courtesy of Gary W. Grimes.) D. *Stylonychia mytilus* normal cells. (Courtesy of Gary W. Grimes.) E. Unidentified ciliate. (Courtesy of E. Small.) These ciliate range in size from about 20 to 100 micrometers.

These multiple kinetids, arranged in rows called *kineties*, can be observed and followed through mitotic division and conjugation. It seems highly likely to us that the cortical phenomena of differentiation so well known for ciliates also occur in other protoctists, animals, and plants, but that the MTOCs and their products are less visible and therefore far more difficult to study. It is in ciliates and their cortical development that the analysis of cell differentiation as a process of microbial community ecology and evolution is best seen.

Applying the principles developed in this book, let us assume that the kinetids directly originated from spirochetes and their attachment sites. Several types of kinetids are shown in figure 51, monokinetids (with a single kinetosome per pattern unit), dikinetids (with two), and polykinetids (with three or more). The major trends in ciliate speciation, then, are not single mutational base changes, although no doubt these mutations have occurred. Ciliate speciation may be reduced to differential growth and morphological change in coevolving microbial communities.

What determines the variation upon which natural selection works in ciliates is, for the most part, the differential growth of the kinetids (kinetosomes, kinetodesmal fibers, and transverse and longitudinal tubules formed by their genetic determinants, MTOCs) relative to the nucleocytoplasm (in other words, population growth of the former spirochetes).

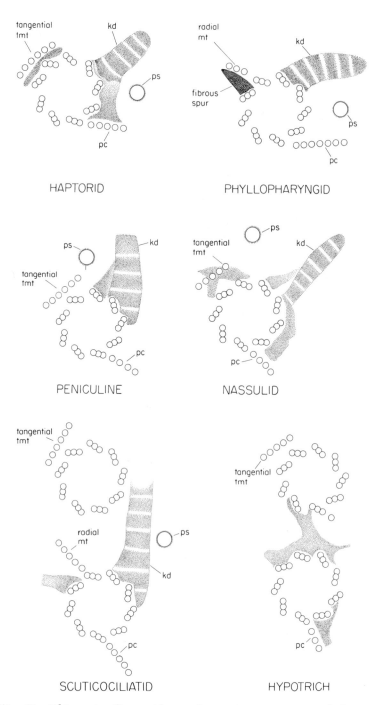

**Fig. 51.** Kinetid types in ciliates with complex cortices. mt = microtubules; tmt = transverse microtubules; pc = postciliary row of microtubules; kd = kinetodesmal fiber; ps = parasomal sac. (Drawing by Laszlo Meszoly.)

Speciation of the kinetids, that is, heritable changes in the morphology of the microbial populations comprising the kinetids, provides a source of variation. These changes, comparable to morphological speciation in free-living spirochetes (Hovind-Hougen, 1978), primarily involve changes in the angles of sets of longitudinal and transverse tubules and kinetodesmal fibers, the appearance of parasomal sacs, and the coevolution of kinetids (for example, the change from mono- to di- to polykinetids) with the rest of the cell (the macronucleocytoplasmic "host," mitochondria, etc.).

A cursory glance at figure 50 suggests immediately that ciliates speciated by changing the numbers and morphological arrangements of the microbial populations that comprised them. Some elegant experiments on the determination of cortical patterns in ciliates have been done by Grimes (1976, 1982) and his colleagues (Grimes et al., 1981) and by de Terra (1983). The results have been difficult to interpret and understand (Grimes, 1980, 1982). Grimes summarizes them in terms of three phenomena: global pattern determination, short-range pattern determination, and developmental assessment. It is our contention that the basis of their understanding lies in the recognition that these processes are fundamentally such processes of microbial ecology and evolution as we have outlined here. Grimes's (1982) summary observation is that the "structural phenotype of these [ciliate] cells is not only inherited true to type sexually and asexually but is also inherited through cystment processes. The number of sets of ciliature, therefore, is determined neither by the nuclear genotype nor by continuity of the visible ciliary structures but rather by some ultrastructurally unidentifiable organization of the cortex." In all cases the individual kinetid and its asymmetry are faithfully retained even though the large structures made up of kinetids (membranelles, cirri) may be arranged and their symmetries altered. The *global patterning* refers to the observations that the number of sets of ciliary structures and the polarity and asymmetry of the overall pattern of those sets are independent of nuclear genes; it is subject to experimental manipulation by surgery and heat shock. The global patterning persists and is inherited faithfully even though all visible traces of the pattern itself disappear during the cystment process. Furthermore, *short-range patterning* depends on the continuous presence of visible ciliary organelles; new kinetids are oriented with respect to old ones in certain morphogenetic sequences (Grimes, 1982). *Developmental assessment* refers to formation, migration, and subsequent proper resorption of ciliary organelles (kinetids, membranelles, and cirri) in the morphogenetic process. This process, which involves the production and destruction of these cortical organelles, clearly indicates the existence of a cellular mechanism of assessment and correction. It ensures that neither hypertrophy nor hypotrophy will ensue and that the number of arrangement of kinetids is optimized for the cell.

We can interpret the phenomena discovered by ciliate biologists—the constancy of the asymmetrical pattern of the individual kinetid and the kinetid's independence from genetic influence of the nucleus and mitochondrial genomes (Beisson and Sonneborn, 1965; Grimes, 1982)—by using the community ecology model. Observations of organelle asymmetry and independence of the nuclei follow directly from the origin of the kinetids from spirochetes. The morphogenesis of each kinetid itself is determined by the residual replicating nucleic acid, the MTOC of spirochete origin, and its interaction with products of the nuclear genome. The fact that laser beam, microsurgical manipulation, and heat shock position kinetids (remnant reproducing spirochetes) suggests that they are still semiautonomous with regard to nuclear genes.

The entire cortical pattern can disappear from view,- during cyst formation, for example, and reappear during the germination of the cyst. This observation is analogous to the dedifferentiation of the A and B symbionts in bugs (homopterans; Schwemmler, 1980); of mitochondria to promitochondria in anaerobically grown yeast (Gillham, 1978); and of plastids to proplastids (Schiff, 1981). It is another example of the dedifferentiation of the MTOC back to ribonucleoprotein from its mature form as a kinetid with undulipodia. The positional information is retained because of the strict and inherited association of MTOCs (which ultimately began as spirochete attachment sites) to the membranes of their hosts. New kinetids are oriented with respect to old ones because they are essentially offspring, products of remnant bacterial cell reproduction, of the old ones.

Developmental assessment is a process in which kinetids regulate their number, position, and interactions. They migrate over the surface forming the cortex, they resorb and reestablish an appropriate number for the size of the host. This phenomenon is well known in grafted stentors and oxytrichs (de Terra, 1983; Grimes, 1980). We interpret this behavior to be the interaction of spirochete populations growing to a maximum number determined by the limits of their environment (their host cells). The grafting of two or more of these ciliates can be accomplished by the elegant manipulations of experts such as Grimes (1976; 1980; 1982), de Terra (1974), Tartar (1961), and Beisson and Sonneborn (1965). In some cases supernumerary kinetids are formed, for example, in the production of new oral membranelles of *Stentor coeruleus* (Tartar, 1961). In order to induce the production of supernumerary kinetids, the part of the cell where thin pigment stripes contact wider ones must be grafted. In these cases the genetic determinants, the replicating MTOCs—apparently residing at the stripe contrast zone—are assumed to have been grafted to the recipient. (These ideas generate an entirely testable prediction: the presence of spirochete-homologous RNA and its polymerase in the transplanted stripe contrast zone and its absence in other parts of the *Stentor* cell, those parts that do not lead to supernumerary kinetids.)

By this reckoning the cortices of ciliates should be considered extremely obvious examples of the relation between microbial population phenomena and development that exist throughout the eukaryotic world. In the rest of the eukaryotic world, however, the phenomenon of the replication, genetic determination, and growth of kinetids is far less obvious. The validity of these ideas can be verified or falsified easily. The verification will come directly from methods found to identify chemically and count the copies of the nucleic acid of spirochetal remnants (the MTOCs) and to localize them within the cell as a function of differentiation, whether on the single or multicell level.

The relationship between the proliferation of kinetids and their genetic determinants and meiosis in ciliates is not obvious. Meiosis evolved independently in ciliates, as it did in several other undulipodiated protoctist lineages: at least heliozoa, foraminifera, labyrinthulids, volvocales, chlorophytes, hypermastigotes, and opalinids (Raikov, 1982). Based on the arguments we have presented here, it is likely that many MTOC genetic determinants, in the form of the original spirochete DNA, have migrated to the micronucleus and that in some cases these code for proteins such as $\alpha$ and $\beta$ tubulin and the proteins of the original spirochete ribosomes. On the other hand, original spirochete gene products—at least, some proteins—must interact with replicating RNA permanently attached at membranous sites in the cytoplasm, that is, cortical RNA. The fact that cortical information is replicated and inherited directly implicates either RNA or DNA; the absence of DNA associated with MTOCs (Younger et al., 1972) implicates replicating, cortical RNA. The concept of replicating, cortically-bound RNA in the MTOC is supported by the direct observations of cortical RNA (Dippell, 1976; Hartman, Puma, and Gurney, 1974). The most likely mechanism of cortical RNA replication is through cortical RNA replicase, which should be actively sought. Such a replicase should not only show homology with some appropriate spirochete RNA or DNA polymerase but should be polarized or positioned in the membrane.

The example of Q$\beta$ is instructive. Q$\beta$ is a bacterial RNA virus that requires only a holoenzyme for its replication. This enzyme has four parts, all of which must be present for RNA replication to occur: the replicase portion coded for by the viral RNA and three proteins from the bacterial host (S1 [ribosomal small subunit 1 protein] and two elongation factors, proteins one of which is sensitive and one insensitive to temperature; these are EFTU and EFTS, respectively).

These three proteins, present in *E. coli*, are likely also to have been present in the gram-negative spirochetal bacterial ancestor to undulipodia. The RNA replication activity responsible for new kinetids, we predict, will be based on a replicase system homologous to ribosomal elongation factor protein components in the relevant spirochetes. MTOCs must minimally be composed of an RNA replicating system (RNA and its polymerase) coding for a MAP, a microtubule associated protein that recognizes tubulins.

## DIFFERENTIATION AS THE NUMBERS GAME

Cell and tissue differentiation are the products of far more than nuclear gene expression in a soluble cytoplasm. Differentiation is fundamentally a manifestation of microbial population biology, what we have called here the "numbers game." That is, cells of individual animals and plants differ from each other because of differential growth of the components comprising them: some have fewer mitochondria and more MTOCs, others more mitochondria and plastids. Each class of organelles has its own genetic systems and autopoietic molecular machinery. Differentiation, then, is the outward manifestation of the differential genome growth and interaction between the nucleocytoplasmic and organellar genomes and their protein synthetic systems.

## DIFFERENTIATION WITHOUT MEIOSIS

Obviously some cell differentiation is possible in the total absence of meiosis. All prokaryotic differentiation occurs without it. Spores, heterocysts, akinetes, and the like are examples. *Fungi imperfecti*, which are considered to be ascomycotes and basidiomycotes that have apparently lost meiosis, seem to have lost simultaneously the ability to differentiate elaborate structures. These cases are particularly evident because the elaborate structures (ascocarps and basidiocarps) are the direct morphological manifestations of the meiotic phases in the life cycle of these fungi. The concept of the obligate relationship between meiosis and tissue and organ differentiation can probably be tested by studies of vegetatively propagated flowering plants. The prediction is clear: meiosis ought still to be obligatory in perhaps nearly half of the 500,000 species that have bypassed mixis (Primack, 1984, pers. com.).

Our final message of this chapter is simply an admonition to reconsider the nature of differentiation and development in eukaryotic organisms. It is a plea to biomathematicians and molecular biologists to concern themselves with intermediary levels of organization between those of macromolecules and organisms. The levels of organization referred to here are those of the eukaryotic cell, but they are equivalent to population growth, physiological and ecological interaction among bacteria, and bacterial evolution—all within the volume bounded by a membrane.

## LOSS OF MEIOTIC SEXUALITY

Even though sexuality may generate in us passions such as ecstasy, jealousy, anxiety, love, and hate, it is not, biologically, an ultimate priority. It is certainly possible, and not against the principles developed here, that methods to bypass sexuality, to omit biparental heritage in the life cycle of people, will be developed by biologists and biochemists. Of course, in case the genetic engineering of human cloning is successful, autopoiesis and reproduction will be retained. Autopoiesis and reproduction are imperatives characteristic of all clones, whether they reproduce in the absence of sexuality or not.

We have seen that sex never appeared at all in many species of bacteria and protoctists. In others (protoctists, fungi, and plants) sex became completely gratuitous. Although it left traces in many species, it disappeared. Vestiges of sexuality, in the form of sterile copulatory embraces or production of eggs capable of development in the absence of fertilization, remind us of the former presence of sexuality in the ancestry of the lineage. In general, it is very difficult to observe directly the process of loss of sexuality. Since the ancestors of many plants, fungi, and animals engaged in well-known sexual practices with morphological correlates, it is inferred that many species have secondarily lost their sexuality, rather than never gained it.

One case of the loss of sexuality has been particularly baffling but well observed. This is the case of the large, blue ciliate *Stentor coeruleus*, a common pond-water protist. Although only a single cell, it is enormous and markedly differentiated (fig. 52A). *Stentor*, which has been under continuous scrutiny since Ehrenberg's pioneer work (1838), seems to be in the process of losing its sexuality.

Each individual *Stentor coeruleus* is capable of a wide range of physiological and behavioral responses. The largest *Stentor coeruleus* measure nearly a millimeter (750 micrometers) long, whereas the smallest, produced by starvation or the experimental induction of continuous mouth regeneration, may be a hundred times smaller (Tartar, 1961). In any culture of stentors in the laboratory, changes in the environmental conditions lead to changes in individual stentors. (For example, drastic changes in the blue pigmentation accompany changes in light conditions; Tartar, 1961; Blumberg et al., 1973.) Although there is a great deal of observable variation in any *Stentor coeruleus* population, no mutants, altered forms that breed true asexually, have ever been reported. Although what maintains variation in stentors is not known,

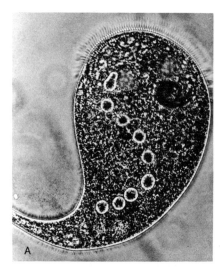
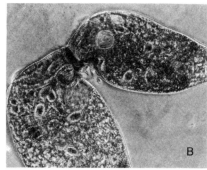

Fig. 52. *Stentor*. A. The protist *Stentor coeruleus*. B. *Stentors* mating.

each stentor has the potential to express any of an entire range of genes specifically in response to environmental changes. All isolates can "regulate," that is, form a whole repertoire of normal stentors (Tartar, 1961). Stentors will occasionally indulge in meiotic sex . . . or try to (fig. 52B)—but both mating partners die within about four days of starting the act of conjugation. Since *Stentor* sex always ends in demise for both partners, we assume that sex, in this species, is on its way to being lost.

Many related ciliates are successfully sexual (Raikov, 1969, 1972). Once useful, the orderly process of meiosis, followed by nucleus swapping and then fertilization, has apparently outlived its purpose in this ciliate. The major way in which *Stentor* populations differentiate is by changes in the number and distribution of the oral and body kinetids. Apparently mixis in the form of conjugation has little to do with this style of differentiation.

Such plants as seedless grapes, oranges, and bananas can be selected for the loss of sexuality quite easily as long as reproduction is ensured by other means, such as farmers. Asexual reproduction works quickly and securely in predictable, human-controlled environments. We human beings provide many such food plants, through agricultural grafting and growing from cuttings, with just such stable environments. The extent to which meiosis is involved directly in differentiation needs to be investigated, but in any case mixis is bypassed in these plants.

In certain organisms sexuality is reserved for only some members of the species. Many social insects have sacrificed mixis: they can no longer reproduce through sex. Termites, no longer able to propagate by pairs, instead depend on special reproductives. The workers forage, feed, and protect the "queens," which mate with the male "kings." Together they bring more sterile soldiers and workers into the world. The determination of sex is by diet and hormone regulation: an insect with a given genetic constitution can give rise to several castes. Obviously in these societies the individual is selected for only as part of the collective. Sterile castes demonstrate how easily loss of mixis can be selected for secondarily, when reproduction is usurped by other means such as that of the royal castes.

Likewise, science fiction and futurology have been full of fanciful possibilities for human societies in which biparental sexuality disappears. The technological procedure of cloning has been the most frequent suggestion. *Brave New World* (Huxley, 1946), a novel based on the application of the totalitarian biology of social insects to human affairs, puts the reproduction of the individual under complete control of the state. No doubt the details of the loss of biparental sexuality differ drastically in many realistic scenarios. Although we are unprepared to offer predictions, we can point to some possible general trends, based on the richness of the variety of sexual behaviors in the past. The success of mechanical (IUD, condoms) and chemical (steroid "pills") methods of birth control has already lessened the frequency of mixis in the human population at large. The Indian government has performed forced sterilizations. China has instituted severe economic penalties on parents of more than a single child. In one possible evolutionary scenario of the future of sex we can imagine this trend continuing until only a tiny fraction of the population is responsible for all human procreation. Coupled with cloning of "desirable" individuals, mixis in human populations could disappear. Homosexual relations provide a clear example of the separation of gender and human sexual behavior from mixis, furthering the trend toward loss of outbreeding, especially in urban technological societies. Even though major Western institutions, such as the Catholic Church, attempt to retain the reproductive traditions of *Homo sapiens* mixis, as in the biblical writ to "be fruitful and multiply," the trends in the educated, densely populated regions of the world are toward a decrease in the frequency of mixis. The idea that enhanced social control (or group coordination) may give a selective advantage to the collective (as it did in the formation of protoctists and animals) supports the idea of future loss of mixis in the descendants of *Homo sapiens*. It is possible to imagine that if mixis in people becomes completely superfluous, evolution will limit the central nervous system feedback needed to provoke the pleasure that leads (presently through sex) to reproduction. Pleasure might, for instance, be experienced as individual members of *"Homo prob-*

*lematicus"* perform the functions necessary to prolong the survival of what-ever larger reproducing entity it is of which they now form a part. The conservative nature of evolution being what it is, central nervous system pleasure might be conserved for other purposes. People might feel pleasure, for instance, as they fulfill cyber-biological roles as "human operators," per-forming maintenance functions that help automata reproduce in space (Evans, 1981).

## SEXUALITY INFERRED FROM THE FOSSIL RECORD

During the course of the history of microbial life on the pre-Phanerozoic Earth there was a significant transition from an atmosphere devoid of mo-lecular oxygen to the current atmosphere, which contains oxygen to a quantity of over 20 percent by volume (Cloud, 1983). Most atmospheric scientists who try to explain the properties of Earth's Archean atmosphere agree that the increase in atmospheric oxygen marked the end of the ultraviolet threat. The production of oxygen led directly to the production of ozone in the upper atmosphere, shielding surface life possibly forever from the lethal effects of ultraviolet light. If this is the case, the recombinatory bacterial protection techniques against ultraviolet light evolved prior to the onset of the Pro-terozoic Eon. The most acceptable date for the beginning of the transition to the oxygen-rich atmosphere is 2,200 million years ago (Cloud, 1974). The oldest evidence for life, far earlier, dates to about 3,500 million years ago (Barghoorn, 1971). We have retained in our cells mechanisms for genetic recombination—that is, the basis of bacterial sexuality—that evolved over 2,000 million years ago and perhaps as long as 3,400 million years ago. We have been using these cut-and-patch mechanisms for other purposes than ultraviolet protection, probably for over three billion years.

The thesis of this book rests, of course, on the truth of our assertion that MTOCs were formerly motile symbionts. If so, their parts were not only used by the bacterial consortium to distribute other genes, but these sorts of illegitimate matings, recombinations between microbes of different ances-tries, were used and are used today in mitosis, meiosis, differentiation, and development. Unfortunately, the cellular events hypothesized here would not have left direct traces in the fossil record. Their consequences, however— for example, the appearance of animals—are indelibly inscribed in the fossil record of life. Encysted structures, interpreted to be early eukaryotes, some quite profoundly decorative, are known from the late Proterozoic and the beginning of the Phanerozoic (Vidal, 1984; fig. 53). These structures are removed from rocks such as shales and sandstones by a maceration technique

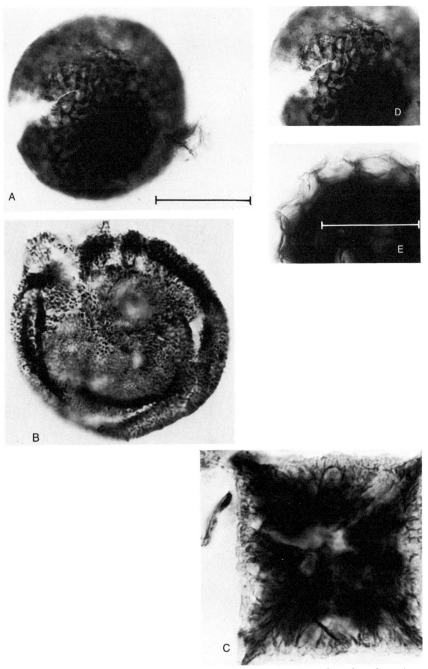

**Fig. 53.** Microfossils of organic-walled organisms interpreted to be algae (eukaryotes). Derived from the Visingsö Group in southern Sweden, the specimens are dated by isotope decay methods to be from 800 million to 700 million years old. All are from acid-resistant residues in macerates from mudstones. A–D, bar = 50 micrometers. E, bar = 10 micrometers. (Courtesy of Gonzalo Vidal.)

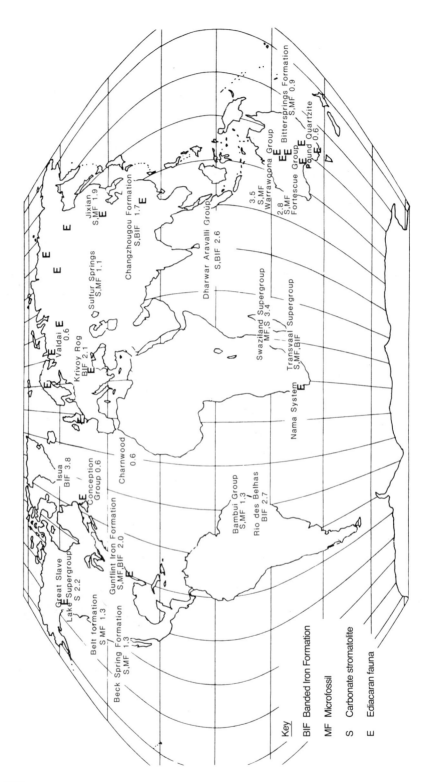

Key

BIF  Banded Iron Formation

MF  Microfossil

S  Carbonate stromatolite

E  Ediacaran fauna

Isua
BIF 3.8

Great Slave
Lake Supergroup
S 2.2

Conception
Group 0.6

Charnwood
0.6

Belt formation
S,MF 1.3

Beck Spring Formation
S,MF 1.3

Gunflint Iron Formation
S,MF,BIF 2.0

Valdai
0.6

Krivoy Rog
BIF 2.1

Sulfur Springs
S,MF 1.1

Jixian
S,MF 1.9

Changzhougou Formation
S,BIF 1.7

Bambui Group
S,MF 1.3

Rio des Belhas
BIF 2.7

Dharwar Aravalli Group
S,BIF 2.6

Swaziland Supergroup
MF,S 3.4

Transvaal Supergroup
S,MF,BIF

Nama System

Bittersprings Formation
S,MF 0.9

Pound Quartzite
0.6

Fortescue Group
S,MF 2.8

Warrawoona Group
S,MF 3.5

involving the dissolution of the rock matrix. The spheroids, which are called "acritarchs," have been interpreted to be remains of early planktonic eukaryotes. Although their true nature is difficult to discern with certainty, they may be providing us with the first fossil evidence for meiotic sexuality. In any case, the complexity and morphological diversity of these acritarchs increase sharply at the beginning of the Phanerozoic Eon (Vidal, 1984). This suggests that the steps we outlined in chapter 11 of the origin of differentiation had already been taken by at least 600 million years ago.

Independent of our interpretation of pre-Phanerozoic acritarchs is other information on the actual date of the origin of meiotic sexuality. The Ediacaran fauna, a set of fossil assemblages of marine invertebrate animals, had well-developed tissues and organs. The presence of these tissues and organs in the fossil record underlines the concept that tissue differentiation, and therefore meiosis, had already evolved by this time. The Ediacaran assemblage of animals has been well documented at nearly two dozen fossil localities all around the world (Glaessner, 1984; fig. 54). Its members may contain representatives from the arthropod, coelenterate, sponge, and annelid phyla, all of which are known from rocks dating from over 700 million years ago. By that time the fundamental patterns of meiosis, fertilization, embryogenesis, and histogenesis were well developed. Sexuality, then, both prokaryotic and meiotic, enjoys an extraordinarily ancient history, one inextricably bound to the history of life.

Sexuality, in all its multiple guises and subtleties, has been an incessant preoccupation of people. Indeed, it is so near our thoughts so frequently that neuter nouns in the Romance languages are prefixed with sexual articles. Poets and scientists have often imbued the universe with their own sexual attributes. It is doubtful that we can agree with the geologist T. C. Chamberlin's claim (1916) that our "planetary system must clearly have had a biparental origin" (pp. 101–02). Nevertheless, with a renewed respect for the

**Fig. 54.** Life in the Pre-Phanerozoic Eon: localities of major evidence for life prior to the appearance of skeletalized animals in the Cambrian period. Numbers indicate age in billions of years. E represents Ediacaran soft-bodied animal fossils, primarily preserved in sandstone. North American finds of Ediacaran fauna include those in North Carolina, southeastern Newfoundland, and the Mackenzie Mountains of northwestern Canada. Eurasian finds include those in Leicestershire, England; Lake Torneträsk, Sweden; northern Russia; western Ural mountains; Turukhansk; Irkutsk (Lake Baikal); northern Siberia; River Maya, China; Longshan, northeastern China; and Yangtze gorge, China. Ediacaran animal remains are also in Namibia in Southwest Africa and probably in Mato Grosso, Brazil, and central Iran. (Drawing by Steven Alexander after Glaessner, 1984.)

origin and ancient history of the complex phenomenon, we can sympathize with Chamberlin's extension of sex to include the abiological union of celestial bodies. In using sex as a metaphor to illuminate processes of planetary formation through collision, Chamberlin shows the intrinsic breadth of the often narrowly conceived and vague notion of sex. Sex should be thought of broadly. The mixing of genes acts at many levels, none of which are mutually exclusive. We hope we have been able to present the circumstantial evidence connected with the various origins of sex in a logical fashion. If the reader feels any closer to the bottom of this evolutionary mystery than when he or she began, we will be satisifed.

# GLOSSARY

ACRITARCH. Late-Proterozoic and early-Phanerozoic spherical walled microfossil thought to be the remains of early eukaryotes.

ACTIN. A class of proteins having molecular weights of about 44,000; involved in the contraction of muscles and many other cell movements.

AEROBE. An organism that lives only in the presence of free oxygen.

AGAMONT. Organism that can reproduce by mitosis but cannot make cells or nuclei that undergo fertilization or meiosis; e.g., apicomplexan schizont.

AKINETE. Propagule of cyanobacteria; generally specialized cell or portion of trichomes capable of separation from parent colony and further growth.

ANAEROBE. Organism that lives only in the absence of atmospheric oxygen.

ANEUPLOIDY. The biologically dangerous state (in animals and plants but not necessarily in protoctists) of having an irregular number of chromosomes (not haploid or diploid).

ANGSTROM (Å). Unit of measurement equal to a millionth of a micrometer, or $10^{-10}$ meter.

ANISOGAMY. The production of gametes that differ in size or form.

APOMICTIC. Formerly sexual (see Table 12).

ARCHEAN. The geological eon beginning with the formation of a solid crust on Earth and continuing to the Proterozoic (from 3,900 until 2,500 million years ago).

ATP. Adenosine triphosphate molecule, a nucleotide of the base adenine with three phosphate groups; used universally in organisms as a source and storage place of metabolic energy.

AUTOGAMY. Self-fertilization, a type of karyogamy; the union of two nuclei both derived from a single parent nucleus.

AUTOPOIESIS. Organismal self-maintenance, a prerequisite to reproduction and thought to have preceded reproduction in evolution; also spelled "autopoesy" or "autopoiesy," from the Greek for "self-making."

AUTOTROPHY. Nutritional mode in which inorganic carbon ($CO_2$) serves as a source of carbon and light or inorganic compounds serve as a source of energy.

AXON. Long, thin extension of a neuron that is underlain by microtubules and that carries impulses away from the cell body.

AXONEME. The 9 + 2 shaft of an undulipodium.

AXOPOD. Long, thin cell structure of protoctists (most conspicuous in the axopodia phylum, such as heliozoans) used for locomotion or feeding; the centers of axopods are made of microtubular shafts.

227

AXOSOME. Fuzzy body where the central pair of microtubules of the axoneme contacts the kinetosome.

BACTERIAL-STYLE SEX. As opposed to regularized cycles of meiosis and fertilization, the free exchange of genes from one entity (usually a virus or bacterium) to another (usually a bacterium); marked by the seemingly indiscriminate amount of nucleic acid, ranging from virtually no DNA to all the donor's genes, passed between the two; bacterial-style sex, including the transfer of plasmids, episomes, and other replicons, need not be confined to bacteria but can occur between organelles within eukaryotes and between bacteria and eukaryotic cells, as in the transfer of the T plasmid of *Agrobacterium* to plant cells.

BACTERIOPHAGE. A replicon covered with a protein coat and hence stable in nature that can infect and grow in bacteria; also called a "phage" or "bacterial virus."

BIOTIC POTENTIAL. Number of organisms that can be produced in a single generation, characteristic of the species. Illustrates tendency of organisms to increase exponentially when their conditions for material growth are satisfied.

BLUE-GREEN ALGA. Old name for a cyanobacterium; oxygenic photosynthetic bacterium.

CANNIBALISM. Eating of conspecifics (members of the same species); this process may have been seminal in the origin of meiotic sex, as it provides a mechanism for cell fusion.

CENTRIOLE. Kinetosome lacking an axoneme; a 9 + 0 structure that forms at each pole in the mitotic spindle during division in most animal cells.

CHLOROPLAST. Green photosynthetic organelle of plants that most likely evolved from endosymbiotic photosynthetic bacteria.

CHROMATID. One of two matching halves of chromosomes found after DNA replication.

CHROMATIN. Long nucleosome-studded DNA and protein that forms chromosomes of eukaryotes.

CILIARY NECKLACE. Pattern of proteinaceous protrusions in membrane at bases of mature undulipodia.

CILIATE. Any of a phylum of protoctists having dimorphic nuclei (macronuclei and micronuclei) and a complex surface layer with characteristic kinetids called a "cortex."

CILIUM (*pl.*, cilia). Undulipodium; generally short 9 + 2 wavy cell protrusion found on eukaryotic cells and used for moving cells or extracellular fluids.

CLIMAX FOREST. A forest that has reached a stage of relative equilibrium with respect to the number and kinds of trees and other organisms living within it.

CLONE. Asexual individual or population derived from a single individual.

COMMUNITY. A unit in nature comprised of populations of organisms of different species living in the same place at the same time; microbial communities are those lacking significant populations of animals and plants.

COMPARATIVE PROTISTOLOGY. The new science of comparing protists and protoctists with the evolution of the nucleus, cell motility, mitosis, meiosis, and so on.

CONJUGATION. In prokaryotes, the transmission of genes from a donor to a recipient cell; in eukaryotes (fungi, red algae, etc.), the fusion of nonundulipodiated gamonts, gametes, or gamete nuclei.

CROSSING OVER. The exchange of portions of homologous chromosomes that occurs during meiotic prophase.

CYCLICAL POPULATION SUCCESSION. An ecological process of community interaction and feedback from a pre-cursor state to a climax, as in the tendency of felled forests to reestablish themselves; the concept that the culmination of this process is reached in the development and differentiation of animals is put forward in this book; in microbial ecology a defect in cyclical population succession is conceptually the same as a birth defect.

CYST. Walled, encapsulated, or dormant roughly spherical structure formed by microorganisms, often in reaction to environmental stress; cysts tend to have greater resistance to desiccation, cold, or heat compared with actively growing forms of the same organism.

CYTOGENETICS. The study of karyotypes (numbers and morphology of chromosomes), usually in an evolutionary context.

CYTOKINESIS. Division of cytoplasm.

DENDRITE. Extension of a neuron that carries impulses toward the cell body; these structures are usually shorter than axons and are underlain by microtubules.

DIFFERENTIAL REPRODUCTION. The concept, central to evolution, that certain organisms produce more offspring and thus are more fit than others with which they may be compared.

DIFFERENTIATION. The process of change in cells, a part of the largely mysterious embryological development from fertilized cell to recognizable individual with distinct tissues and organs.

DIKARYON. A cell with two differently derived nuclei sharing a common cytoplasm.

DIKARYOSIS. The process of making or the state of being a dikaryon.

DINOFLAGELLATE. A common name for a dinomastigote, one of a phylum of undulipodiated microbes important in tracing eukaryote evolution

because of their idiosyncratic nuclear and mitotic apparatus.

DIPLOIDY. 2N state in which there are two sets of chromosomes in the nuclei of eukaryotic cells.

DUPLICATION. Complementary doubling of DNA or RNA molecules; appearance of two new chromosomes from chromatids after reproduction of the centromere in metaphase of mitosis.

DYNEIN. The microtubule-associated protein with ATPase activity comprising the "arms" on the outer doublet tubules of undulipodia.

ECOSYSTEM. A unit in nature, comprised of communities in which the biologically important chemical elements (C, N, S, P, and so on) cycle completely; these elements cycle more rapidly within an ecosystem than between ecosystems.

ECTOMYCORRHIZAE. Symbiotic association generally between basidiomycote fungi and the roots of woody plants; the fungal growth covers the outside of the roots and metabolic changes occur.

EMBRYOGENESIS. The development of embryos.

ENDOCYTOBIOLOGY. The study of cells within cells (term coined by Schwemmler; see Schwemmler, 1980).

ENDOCYTOSIS. General term for "cell eating"; phagocytosis of particulate matter in a cell that may or may not bear pseudopods.

ENDONUCLEASE. An enzyme that cuts into DNA at any place along its length except at its ends.

EPISOME. Small replicon that may replicate faster than host genetic material and later become incorporated into the genophore.

EUKARYOTE. Cell (or organism of cells) having a membrane-bounded nucleus; by hypothesis derived from co-evolved microbial communities.

EUPLOIDY. The state of having a complete set of chromosomes.

EXCISION REPAIR. A DNA repair system that uses a healthy, undamaged strand of DNA to replace a damaged one.

EXHABITANTS. Outside partners in a symbiosis; i.e., hosts facing the environment and containing symbionts.

EXOCYTOSIS. A process of cellular waste removal or particle secretion involving movement of material from the interior of the cell through the membrane (reverse of endocytosis).

EXTRANUCLEAR ORGANELLES. Microtubule organizing centers (MTOCs) external to the nucleus. Often refers to hypermastigotes.

FEMALE. Convenient term for the gamont in anisogamous species that produces gametes that are relatively large, immotile, and few in number.

FERTILIZATION. Fusion of gamonts, gametes, or gamete nuclei; the eukaryotic complement to meiosis that together with it forms the basis of sexual reproduction, the doubling of chromosome numbers in cell fusion.

F-FACTORS. Fertility factors of bacteria. Type of small replicon.

GAMETE. Haploid reproductive cell whose nucleus fuses with that of another gamete during fertilization.

GAMETIC MEIOSIS. The type of meiosis that precedes the formation of gametes or gamete nuclei; characteristic of some protoctists and all animals.

GAMETOGAMY. Fusion of gametes (cells or nuclei). First step in fertilization.

GAMETOPHYTE. Type of gamont. Haploid spore-producing individual or generation of a plant that has generations of alternating ploidy; e.g., mosses and ferns.

GAMMA PARTICLE. Small, symbiont-like organelle that goes through peculiar life-cycle changes within *Blastocladiella* and that is the size of a large virus.

GAMONT. Eukaryotic organism capable of producing gametes or entering sexual encounter by producing cells with haploid nuclei.

GAMONTOGAMY. Aggregation or union of gamonts. Mating.

GENDER. The set of secondary sexual characteristics generally correlating with mating types.

GENOME. The total DNA makeup of an organism.

GENOPHORE. Bacterial large DNA replicon (bacterial "chromosome").

GERM LINE. In animals and plants, the set of "immortal" cells, such as the oocytes and their predecessors that become eggs, and spermatocytes and their predecessors destined to become sperm or gamete nuclei.

GERM PLASM. Germ line (non-soma or nonsomatic cells).

GOLGI BODY. Intracytoplasmic membranous structure made of flattened saccules, often stacked in parallel arrays, and vesicles; also called "dictyosome," "Golgi complex," "parabasal body."

HAPLOIDY. 1N state in which there is a single set of chromosomes in the nuclei of eukaryotic cells.

HERMAPHRODITISM. The condition in animals of having both male and female sexual organs; called "monoeciousness" in plants (see Table 12).

HETEROCYST. Cyanobacterial structure made of specialized thick-walled cells capable of nitrogen fixation.

HETEROGENOMIC. Having two genomes of different evolutionary origins in a single cell or organism; descriptive of genomes of permanent symbionts.

HETEROKARYON. Kind of dikaryon; fungal hypha containing two genetically distinguishable nuclei.

HETEROLOGOUS GENOME. Total DNA makeup characterized by the presence of two or more genomes of different evolutionary origins.

HETEROTROPHY. Nutritional mode of organisms that gain both carbon and energy from organic compounds (ultimately produced by autotrophs).

HISTONE. One of a class of lysine- and arginine-rich positively charged chromosomal proteins that bind to DNA.

HOMEORRHESIS. Self-regulation around moving set points, as in the embryology of animals such that limbs and organs continually grow but are kept in proportion.

HOMEOSTASIS. Self-regulation of temperature, chemical composition, and acidity around a value called a "set point," as in the regulation of body temperature in mammals.

HYPERMASTIGOTE. Any of several motile parabasalid heterotrophic protists, some of which bear thousands of undulipodia.

HYPHA. Tubular filament that forms the units of structure in fungi.

IMMUNOGLOBULIN SYSTEM. In vertebrates, the production of specific antibodies capable of binding and removing foreign substances (antigens); antibodies are proteins belonging to the class of immunoglobulins.

INBREEDING. Incrossing; the formation by sexual reproduction of a new individual from genetically related parents.

ISOGAMY. The production of compatible gametes that are equal in size and form.

KARYOGAMY. Fusion of nuclei, usually in fertilization; the nuclei are sometimes called "pronuclei" or "sexual nuclei."

KARYOKINESIS. Nuclear cell division.

KARYOTYPE. The trait of a cell nucleus with respect to chromosome number, form, and size and to position of centromere (points of spindle attachment).

KARYOTYPIC FISSIONING. Spontaneous process of dividing of large into small chromosomes via an extra round of centromeric replication that results in doubling the number of chromosomes in a single step; the process is thought to be important in mammalian evolution.

KINASE. A class of enzymes including, e.g., those that convert an inactive protein precursor (zymogen) into an enzyme, usually involving transfer of a phosphate group.

KINETID. Elementary repeating unit in all undulipodiated cells that consists of a kinetosome (or kinetosomes) and associated organelles, such as undulipodia, striated fibers, and sheets of microtubules.

KINETOCHORE. Centromere; the proteinaceous and nucleic-acid site of chromatid attachment to microtubular spindle fibers during mitosis and meiosis.

KINETOSOME. 9 + 0 organelle at the base of all undulipodia; centriole from which axoneme emerges; formerly called basal body.

LIGASE. A class of enzymes that splice together separate lengths of DNA.

LIGATION. Linking, as in patching of DNA to integrate spliced genes.

LYSOGENY. Phage burst; the phenomenon in which bacteriophages reproduce inside and burst out of bacterial cells.

MACRONUCLEUS. Diploid or higher-ploid large "physiological" nucleus of ciliates that controls the organism's phenotype by the synthesis of messenger RNA; macronuclei are formed by differentiation of micronuclei.

MAINTENANCE OF SEX.    The continuation of sex in sexually reproducing organisms (attributed to natural selection by those who use the term).

MALE.    Convenient term for the gamont in anisogamous species that produces gametes that are relatively small, motile, and numerous.

MAPs.    See    microtubule-associated proteins.

MASTIGOTE.    Undulipodiated    cell, usually motile.

MATING TYPE.    Attribute of gamonts of a species unable to fuse (mate, conjugate) with members of the same type under optimal conditions for mating; the number of different mating types varies from two to nearly 80,000 in a given species.

MEIOSIS.    The reciprocal process of fertilization in which diploid or 2N cells are reduced to haploid or 1N cells (such as eggs, sperm, spores).

MEIOTIC REDUCTION    See meiosis.

MEIOTIC SEX.    The life-cycle combination of meiosis and fertilization that characterizes sexual eukaryotes.

MICROBIAL    COMMUNITY.    Populations of microorganisms, members of different species, found in the same place at the same time.

MICROBIAL ECOLOGY.    The study of microbes and their interaction with each other and the environment.

MICRONUCLEUS.    The small diploid nucleus of ciliates capable of production of more micronuclei by mitosis or of development into a higher-ploid macronucleus; micronuclei are exchanged during sexual processes in ciliates.

MICROTUBULES.    Hollow cylindrical structure the walls of which are composed of the tubulin proteins (α tubulin, β tubulin) and associated with other proteins (MAPs); about 2.5 nanometers (25Å) in diameter.

MICROTUBULE-ASSOCIATED    PROTEINS (MAPs).    A class of proteins, including dynein, which copurify with tubulins; several are high molecular weight relative to tubulin.

MICROTUBULE    ORGANIZING    CENTER (MTOC).    Site in a cell always associated with the appearance and organization of microtubules, such as in the development of the axoneme from the kinetosome in an undulipodium; the smallest are occasionally seen under the electron microscope as fuzzy spots, or their existence may be inferred from appearance of microtubule-based organelles; among the largest are heliozoan MTOCs.

MICROTUBULE    PROTEIN.    Tubulin; any of a class of proteins comprising microtubules and having molecular weights of about 50,000.

MITOCHONDRION.    Oxygen-respiring organelle found in nearly all eukaryotic cells, including those of all plants and animals; thought to be derived from Proterozoic aerobic bacteria.

MITOSIS.    Cell division of most eukaryotes consisting of a doubling of chromosomes followed by their segregation and deployment to offspring cells; mitosis accounts for the reproduction of protoctists and the growth of plants, fungi, and animals.

MITOTIC APPARATUS.    Mitotic spindle; the double-coned structure composed of microtubules that form during mitosis and meiosis and is responsible for chromosome movement into offspring cells.

MITOTIC OPTION.    "Choice" in lineages of animal cells either to reproduce by mitosis or to differentiate undulipodia.

MIXIS.    Production of a single individual from two parents by way of fertilization occurring at level of fused cells or individuals.

MONOECIOUSNESS. The simultaneous presence of male and female sex organs in a plant; in animals, called "hermaphroditism."

MONOMER. Single molecule in a chain of related molecules; e.g., amino acids are the monomers of protein polymers, nucleotides are the monomers comprising nucleic acid polymers.

MTOC. *See* microtubule organizing center.

MULTIPLE FISSION. Cytokinesis following earlier karyokineses and resulting in the simultaneous production by a cell of several offspring cells.

MYOSIN. A large protein showing ATPase enzymatic activity involved in the contraction of muscles and many other cell motility processes.

NAOs. Nucleus-associated organelle; a type of MTOC; also called "extranuclear division center."

9 + 2. Axoneme structure; descriptive of the arrangement of nine triplets (9) of microtubules surrounding a central pair (2); shaft of undulipodium.

9 + 0. Kinetosome structure; descriptive of the arrangement of nine triplets (9) of microtubules and none (0) in the middle; characteristic of kinetosomes at the bases of all undulipodia.

NUCLEOID. Genophore; the large replicon of DNA of a bacterium as visualized in the electron microscope.

NUCLEOSOME. Beadlike form of DNA-protein organization in chromatin.

NUCLEOTIDE. Any of a class of compounds made of nitrogenous bases and pentose sugars (ribose, deoxyribose) with phosphates attached; examples include adenosine triphosphate, thymidine monophosphate.

OC THALLUS. Ordinary colorless thallus of *Blastocladiella;* the most common conspicuous trophic form in the life cycle of this chytrid.

1N. Haploid; descriptive of eukaryotic organisms having one set of chromosomes in the nuclei of their cell(s).

ONE-STEP MEIOSIS. Meiosis resulting from a single reduction division of the nucleus that converts a diploid to a haploid cell.

OUTCROSSING. The sexual reproduction by mating of two organisms no more closely related to each other than to any other members of the population at large.

PARASEXUALITY. Any process except standard meiosis and fertilization (and bacterial-style sex) that produces an individual from more than a single parent.

PARASOMAL SAC. Small membrane-lined depression in the cortex of ciliates.

PARENT. Individual that provides the genetic material for offspring, whether viruses, bacteria, eukaryotic cells, or organisms.

PARTHENOGENESIS. Development of an organism from eggs or macrogametes in the absence of mixis (see Table 12).

PERICENTRIOLAR DENSE BODY. Tiny amorphous structure seen with the electron microscope to be associated with 9 + 0 centrioles.

PHAGOCYTOSIS. "Cell eating"; a kind of endocytosis that involves the formation of pseudopods; characteristic of amoebae, lymphocytes, etc.

PHANEROZOIC EON. The modern eon of Earth history, beginning 570 million years ago with the appearance of Cambrian hard-bodied fossil forms and lasting until the present.

PHOTOREACTIVATION. Enzymatic DNA repair that occurs when DNA-damaged bacteria are exposed to light.

PILUS (*pl.*, pili). Tiny tubelike hair-shaped structure on the outer cell wall surfaces of bacteria that may be directly involved in transfer of DNA from cell to cell.

PINOCYTOSIS. "Cell drinking"; a kind of endocytosis that involves the formation of long channels in the cytoplasm for taking up protein-sized particles visible only with the electron microscope.

PLASMID. DNA not coated with protein and capable of replication at different rates relative to genophores or chromosomes residing in the same cell; a kind of small replicon.

PLASTID. Photosynthetic organelle within plant and algal cells thought to be of polyphyletic endosymbiotic origin from photosynthetic bacteria.

PLOIDY. Number of sets of chromosomes.

POLYMER. Any molecule composed of a long chain of monomers.

POLYMERASE. An enzyme that synthesizes polymers by connecting monomers.

POLYPHYLY. Multiple evolutionary origins.

POLYSACCHARIDE. Carbohydrate molecule composed of a long chain of sugar monomers; e.g., starch, cellulose.

POPULATION. A group of organisms belonging to the same species and living in the same place at the same time.

PRE-PHANEROZOIC. The period of Earth history predating the appearance of hard-bodied animals in the Cambrian period 570 million years ago; the Hadean (?–3,900 m.y.a.), the Archean (3,900–2,500 m.y.a.), and the Proterozoic (2,500–600 m.y.a.) eons comprise the Pre-Phanerozoic.

PROKARYOTE. Cell or multicellular organism lacking membrane-bounded nuclei; e.g., a bacterium.

PROMISCUOUS DNA. DNA associated with one sort of organelle (e.g., chloroplasts) found in another class of organelle (e.g., nucleus or mitochondria).

PROTEROZOIC. The eon of Earth history stretching from 2,500 million years ago to the Cambrian period at the beginning of the modern Phanerozoic Eon.

PROTIST. Unicellular protoctist.

PROTISTAN NUCLEAR CYTOLOGY. The study of protist karyotypes and nucleocytoplasmic relations in protoctists (see Raikov, 1982).

PROTOCTIST. A member of the kingdom Protoctista, unicellular (protists) and multicellular eukaryotic microorganisms and their descendants excluded from animals (K. Animalia), plants (K. Plantae), and fungi (K. Fungi).

PROTOEUKARYOTE. Organism assumed to be representative of eukaryote ancestors.

PROTOMITOCHONDRION. Putative free-living ancestor to mitochondrion; thought to be *Paracoccus-*, *Bdellovibrio-*, or *Rhodopseudomonas*-like bacterium.

PROTONUCLEOCYTOPLASM. Putative free-living, largely anaerobic host of eukaryotes; thought to be *Thermoplasma*-like archaebacteria.

PSEUDOGENE. One of a class of apparently nonfunctional DNA that, on the basis of its nucleotide sequences, can be related to recognizable structural genes.

RADICAL. Chemical term, abbreviated "R," for electrically charged cluster of elements that is electrically unstable in that it is highly reactive; carbon-hydrogen radicals are often labeled "R" when attention is to be drawn to the rest of the molecule.

RAMET. Asexual or asexually produced generative organ or organism (animals).

RECOMBINANT DNA. Genetic recombination; i.e., sex on the nucleic acid level.

REDUCTION DIVISION. Meiosis.

REPLICATION. Copying of genetic material (DNA or RNA).

REPLICON. Naturally replicating entity, such as a bacterial genophore, a plasmid, or a virus.

REPRODUCTION. The process that augments the number of cells or organisms; not to be confused with sex, which is any process that recombines genes (DNA) in an individual cell or organism from more than a single source.

RESISTANT SPORANGIUM. Large heavy-walled structure in the chytrid *Blastocladiella* that is desiccation-resistant and that potentially contains and releases zoospores.

RNP. Ribonucleoprotein.

ROOT FIBERS. Striated rootlet, kinetodesmal fiber, or other kinetid structure.

SCHIZONT. An agamont, especially in apicomplexan parent cells, that produces many offspring (called "merozoites").

SEME. Complex trait of clear selective advantage, and therefore evolutionary importance, resulting from evolution of an interacting set of genes; e.g., wings, nitrogen fixation.

SEX. Any process that recombines genes (DNA) in an individual cell or organism from more than a single source; sex may occur at the nucleic acid, nuclear, cytoplasmic, and other levels; *see also* maintenance of sex.

SEX CELL. Gamete; cell produced by gamonts, such as sperm, eggs, or other haploid, meiotically derived cells, that have the potential to become fertilized; products of germ line.

SEXUAL DIMORPHISM. The tendency within a species of members of different genders (male and female) to develop differing size and form; *polymorphism* refers to the existence of three or more forms.

SLUG. The group-structure stage in the life of slime molds in which individual amoebae come together.

SOS REPAIR. A DNA repair system that allows damaged DNA with mispaired bases to replicate, leading to the production of many mutants.

SPIROCHETAL SECRET AGENT. Putative remnant of symbiotic spirochetes in modern cells; examples include the mitotic spindle, kinetochores, kinetosomes, undulipodia, MTOCs, and other structures.

SOMA. Somatic cells, not germ line cells, that is, those nuclei or body cells of organisms displaying programmed death, those unable to become gametes or gamete nuclei.

SPIROCHETE. Corkscrew-shaped heterotrophic bacterium similar in morphology to spirilla except that the flagella of spirochetes are inside the outer membrane of the gram-negative cell wall; many spirochetes are extremely rapid swimmers, and they proliferate especially in anaerobic environments, such as muds and the guts of animals.

SPIROCHETE HYPOTHESIS. The hypothesis that undulipodia originated from motile anaerobic spirochetes which entered, reproduced in, and became symbiotic with protoeukaryotic cells. Spirochetes along the way lost the ability to reproduce outside the cytoplasm as they differentiated into structures at the basis of eukaryotic cell motility, such as MTOCs.

SPORANGIOSPORE. A propagule of

protoctists and fungi borne in spore cases (sporangiophores).

SPORE.    Propagule; a small, highly polyphyletic, often desiccation- or heat-resistant structure capable of development into a mature organism.

SUPERORGANISM.    Precisely integrated ecological community.

SYMBIONT.    An organism that lives with another of a distinct species or kind for most of the life cycle of both.

SYMBIOSIS.    An association between members of different species in which the partners live in physical contact for most of the life cycle of both.

SYNAPTONEMAL COMPLEX.    Complicated linear proteinaceous structure that forms between homologous chromosomes in meiotic prophase as well as at other times.

SYNGAMY.    Fusion of gametes.

TETRAPLOIDY.    The condition, common in plant cells, of having four sets of chromosomes.

THOUGHT COLLECTIVE.    Group of persons having a socially conditioned style of thought; e.g., university botanists (see Fleck, 1979).

THOUGHT STYLE.    A focused mode of thinking or "readiness for directed perception" that organizes data in a certain way at the expense of new or alternative explanations; e.g., the idea that sickness is an attack or invasion leads to "wars on disease" (see Fleck, 1979).

THYMINE DIMER.    A "knot" of thymine in which, instead of attaching to cytosine as in the ordinary replication of a DNA molecule, a thymine molecule binds to a second thymine.

TOTIPOTENT.    As applied to a single cell, capable of development along any of the lines inherently possible to organisms of its kind and capable of regenerating a normal organism.

TROPHOZOITE.    The trophic or growing and feeding life-cycle stage (in contrast to an encysted, dormant, or propagule stage).

TUBULIN.    See microtubule protein.

2N.    Diploid; descriptive of eukaryotic organisms having two sets of chromosomes in the nuclei of their cell(s).

TWO-STEP MEIOSIS.    Meiosis resulting from two nuclear divisions that reduce the 2N diploid cell to the 1N haploid cell; the classical meiosis of metazoans in which only one round of chromosome synthesis occurs in two nuclear divisions.

UNDULIPODIUM.    9 + 2 eukaryotic organelle (confusingly referred to as a "eukaryotic flagellum" when long and few per cell and as a "cilium" when short and many per cell); it is agreed that all undulipodia from sperm tails to protist cilia, because of their 9 + 2 ultrastructure, have a common origin (see spirochete hypothesis).

VIRION.    Mature bacteriophage particle, including the small replicon and its protein coverings.

ZOOSPORES.    Undulipodiated propagule.

ZYGOSPORE.    Large multinucleated resistant structure that results from the conjugation of fungal hyphae (in zygomycotes, a fungal phylum, or zygomycetes, a fungal class).

ZYGOTE.    Diploid nucleus or cell produced by the syngamic fusion of two haploid cells and destined to become a new organism.

ZYGOTIC MEIOSIS.    The process in which meiosis occurs immediately after the diploid zygote forms; characteristic of some protoctists and all fungi.

# REFERENCES

Abelson, J. 1982. Self-splicing RNA. *Nature* 300:400.

Adoutte, A.; Claisse, M.; and Cance, J. 1984. Tubulin evolution: An electrophoretic and immunological analysis. *Origins of Life* 13:177–82.

Allfrey, V. G.; Bautz, E. K. F.; McCarthy, B. J.; Schimpke, R. T.; and Tissieres, A., eds. 1976. *Organization and expression of chromosomes.* Life Sciences Research Report 4. Berlin: Dahlem Konferenzen.

Ammermann, D. 1973. Cell development and differentiation in the ciliate *Stylonychia mytilus.* In *Cell cycle in development and differentiation,* ed. M. Balls and F. S. Billett, 51–60. Cambridge: Cambridge University Press.

Anker, P.; Stroun, M.; Gahan, P.; Rossier, A.; and Greppin, H. 1971. Natural release of bacterial nucleic acids in plant cells and crown gall induction. In *Informative molecules in biological systems,* ed. L. G. H. Ledoux, 193–98. Amsterdam: North-Holland.

Atema, J. 1973. Microtubule theory of sensory transduction. *Journal of Theoretical Biology* 38:181–90.

Avery, O. T.; Macleod, C. M.; and McCarty, M. 1944. Studies on the chemical nature of the substance inducing transformation of pneumococcal types. Induction of transformation by a desoxyribonucleic acid fraction isolated from *Pneumococcus* type III. *Journal of Experimental Medicine* 79:137–58.

Bardele, C. F. 1977a. Comparative study of axopodial microtubule patterns and possible mechanisms of pattern control in the centrohelidian heliozoa *Acanthocystis, Raphidiophrys* and *Heterophrys. Journal of Cell Science* 25:205–32.

———. 1977b. Organization and control of microtubule pattern in centrohelidian heliozoa. *Journal of Protozoology* 24:9–14.

———. 1981. Function and phylogenetic aspects of the ciliary membrane: A comparative freeze-fracture study. *BioSystems* 14:403–21.

———. 1983. Comparative freeze-fracture study of the ciliary membrane of protists and invertebrates in relation to phylogeny. *Journal of Submicroscopic Cytology* 15:263–67.

Barghoorn, E. S. 1971. The oldest fossils. *Scientific American* 224:30–42.

Barnett, A. 1966. A circadian rhythm of mating type reversals in *Paramecium multi-micronucleatum. Journal of Cell Physiology* 67:239–70.

Barr, D. J. S. 1978. The flagellar apparatus in the Chytridiales. *Canadian Journal of Botany* 56:887–900.

———. 1981. The phylogenetic and taxonomic implications of flagellar rootlet morphology among zoosporic fungi. *BioSystems* 14:359–70.

———. 1984. Cytological variation in zoospores of *Spizellomyces (Chytridiomycetes). Canadian Journal of Botany* 62:1201–08.

Beale, G. 1954. *The genetics of* Paramecium. Cambridge: Cambridge University Press.

Beisson, J., and Sonneborn, T. M. 1965. Cytoplasmic inheritance of the organization of the cell cortex in *Paramecium aurelia. Proceedings of the National Academy of Science, USA* 53:275–82.

Belar, K. 1926. *Der Formwechsel des Protistenkerne.* Jena: G. Fischer.

Bell, Graham. 1982. *The masterpiece of nature: The evolution and genetics of sexuality.* Berkeley and Los Angeles: University of California Press.

Blumberg, S.; Honjo, S.; Otaka, T.; Antanavage, J.; Banerjee, S.; and Margulis, L. 1973. Induced reversible pigment alteration in *Stentor coeruleus. Transactions of the American Microscopical Society* 92:557–69.

Bogorad, L. 1975. Evolution of organelles and eukaryotic genomes. *Science* 181:891–98.

Bradbury, E. M.; Maclean, N.; and Mathews, H. R. 1981. *DNA, chromatin and chromosomes.* New York: John Wiley and Sons.

Brinkley, B. R., and Stubblefield, E. 1970. Ultrastructure and interaction of the kinetochore and centriole in mitosis and meiosis. *Advances in Cell Biology* 1:119–85.

Brown, S.; Margulis, L.; Ibarra, S.; and Siqueiros, D. 1985. Desiccation resistance and contamination as mechanisms of gaia. *BioSystems* 17:337–60.

Brugerolle, G., and Mignot, J.-P. 1984. The cell characters of two helioflagellates related to the centrohelidian lineage: *Dimorpha* and *Tetradimorpha. Origins of Life* 13:305–14.

Bruns, P., and Gorovsky, M., eds. 1984. *Ciliate molecular genetics. Abstracts of papers.* Cold Spring Harbor, NY: Cold Spring Harbor Laboratory.

Buchanan, R. E., and Gibbons, N. E., eds. 1974. *Bergey's Manual of Determinative Bacteriology.* 8th ed. Baltimore: Williams and Wilkins.

Bulmer, M. G. 1982. Cyclical parthenogenesis and the cost of sex. *Journal of Theoretical Biology* 94:197–207.

Busch, H., ed. 1978. *Chromatin, Part C.* Vol. 6 of *The cell nucleus.* New York: Academic Press.

Buss, Leo. 1983a. Evolution, development and the units of selection. *Proceedings of the National Academy of Sciences, USA* 80:1387–91.

———. 1983b. Somatic variation and evolution. *Paleobiology* 9:12–16.

———. n.d. The uniqueness of the individual revisited. In *Population biology and evolution of clonal organisms,* ed. J. B. C. Jackson, L. W. Buss, and R. E. Cook. New Haven and London: Yale University Press (forthcoming).

Calarco-Gillam, P. D.; Siebert, M. C.; Habble, R.; Mitchison, T.; and Kirschner, M. 1983. Centrosome development in early mouse embryos as defined by an autoantibody against pericentriolar material. *Cell* 35:621–29.

Calder, N. 1983. *Timescale: An atlas of the fourth dimension.* New York: Viking Press.

Cantino, E. C. 1966. Morphogenesis in aquatic fungi. In *The fungi: An advanced treatise,* ed. G. C. Ainsworth and A. S. Sussman. New York: Academic Press.

Cantino, E. C., and Mills, G. L. 1979. The gamma particle in *Blastocladielli emersonii:* What is it? In *Viruses and plasmids in fungi,* ed. P. Lemke, 441–84. New York: Marcel Dekker.

———. 1983. The Blastocladialean gamma particle: Once viral endosymbiont (?), now "chitosome" progenitor. In *Fungal differentiation,* ed. J. E. Smith, 175–209. New York: Marcel Dekker.

Canuto, V. M.; Levine, J. S.; Augustsson, T. R.; and Imhoff, C. L. 1982. UV radiation from the young Sun and oxygen and ozone levels in the prebiological palaeoatmosphere. *Nature* 296:816–20.

Cech, T. R. 1983. RNA splicing: Three themes with variations. *Cell* 34:713–16.

Chamberlin, T. C. 1916. *The origin of the Earth.* Chicago: University of Chicago Press.

Cleveland, L. R. 1935. The cell and its role in mitosis as seen in living cells. *Science* 81:597–600.

———. 1947. The origin and evolution of meiosis. *Science* 105:287–88.

———. 1949. The whole life cycle of chromosomes and their coiling system. *Transactions of the American Philosophical Society* 39:1–100.

———. 1953. Studies on chromosomes and nuclear division I–IV. *Transactions of the American Philosophical Society* 43:809–69.

———. 1956. Cell division without chromatin in *Trichonympha* and *Barbulanympha*. *Journal of Protozoology* 3:78–83.

———. 1957. Types and life cycles of centrioles of flagellates. *Journal of Protozoology* 4:230–40.

———. 1963. Function of flagellate and other centrioles in cell replication. In *The cell in mitosis*, ed. L. Levine, 3–53. New York: Academic Press.

———. n.d. [1950s]. *Flagellates of termites.* 3 pts. 16mm. Department of Zoology, University of Massachusetts, Amherst.

———. n.d. [1950s]. *Gametogenesis and fertilization in* Trichonympha. 16mm. Department of Zoology, University of Massachusetts, Amherst.

———. n.d. (1950s]. *Sexuality and other features of the flagellates of* Cryptocercus. 16mm. Department of Zoology, University of Massachusetts, Amherst.

Clewell, D. B. 1985. Sex pheromones and conjugation in streptococci. In *The origin and evolution of sex*, ed. H. O. Halvorson and A. Monroy. New York: Alan R. Liss.

Cloud, P. E., Jr. 1974. Evolution of ecosystems. *American Scientist* 62:54–66.

———. 1983. The biosphere. *Scientific American* 249:176–89.

Cole, C. J. 1984. Unisexual lizards. *Scientific American* 250:94–100.

Corliss, J. O. 1979. *The ciliated Protozoa.* 2d ed. New York and London: Pergamon Press.

———. 1984. The kingdom Protista and its forty-five phyla. *BioSystems* 17:87–126.

Crick, F. H. C. 1981. *Life itself.* New York: Viking Press.

Crocker, T. T., and Dirksen, E. 1966. Cilia development in fetal rat tracheal epithelium. *Journal de Microscopie* 5:629–44.

Daniels, E. W., and Breyer, E. P. 1967. Ultrastructure of the giant amoeba *Pelomyxa palustris*. *Journal of Protozoology* 14:167–79.

Darwin, C. 1859. *Origin of species by means of natural selection.* See *The illustrated origin of species*, abridged and introduced by Richard E. Leakey. New York: Hill and Wang, 1982.

Davis, B. D.; Dulbecco, R.; Eisen, H. N.; and Ginsberg, H. S. 1980. *Microbiology.* 3d ed. New York: Harper and Row.

Dawkins, R. 1976. *The selfish gene.* Oxford: Oxford University Press.

———. 1982. *The extended phenotype.* San Francisco: W. H. Freeman.

Day, P. R. 1974. *The genetics of host-parasite interaction.* San Francisco: W. H. Freeman.

Day, W. 1984. *Genesis on planet Earth.* New Haven and London: Yale University Press.

de Beer, Gavin. 1972. *Adaptation.* Oxford Biology Readers. Oxford: Oxford University Press.

De Long, R. 1983. A unified concept for microbial genetics. *Journal of Theoretical Biology* 103:163–65.

Desportes, I. 1984. The Paramyxea Levine 1979: An original example of evolution towards multicellularity. *Origins of Life* 13:343–52.

de Terra, N. 1974. Cortical control of cell division. *Science* 184:530–37.

————. 1983. Microsurgical experiments in *Stentor*. Personal communication.

D'Herelle, F. 1926. *The bacteriophage and its behavior*. Baltimore: Williams and Wilkins.

Dingle, A. D. 1979. Cellular and environmental variables determining the numbers of flagella in temperature-shocked *Naeglaria*. *Journal of Protozoology* 26:604–12.

Dippell, R. V. 1968. The development of basal bodies in *Paramecium*. *Proceedings of the National Academy of Sciences, USA* 61:461–68.

————. 1976. The effect of nuclease and protease digestion on the ultrastructure of *Paramecium* basal bodies. *Journal of Cell Biology* 69:622–37.

Dodson, E. 1979. Crossing the eucaryote-procaryote border: Endosymbiotic or continuous development? *Canadian Journal of Microbiology* 25:265–81.

Doolittle, W. H. F., and Sapienza, C. 1980. Selfish genes: The phenotype paradigm and genome evolution. *Nature* 284:601–03.

Dougherty, E. C. 1955. Comparative evolution and the origin of sexuality. *Systematic Zoology* 4:145–70.

Dustin, P. 1978. *Microtubules*. Berlin, Heidelberg, and New York: Springer-Verlag.

————. 1983. Les microtubules et leurs fonctions. Acquisitions recentes. *Treballs de la Societat Catalana de Biologia* 35:15–63.

Dutcher, S. K. 1984. *Chlamydomonas* kinetosome mutants [personal communication]. Department of Cellular, Molecular, and Developmental Biology, University of Colorado, Boulder.

Dyer, Betsey Dexter. 1984. Protoctists from the microbial mat at Laguna Figueroa, Mexico. Ph.D. diss., Boston University Graduate School.

Edstrom, J.-E. 1976. Meiotic versus somatic transcription with special reference to diptera. In *Organization and expression of chromosomes*, ed. V. G. Allfrey, E. K. F. Bautz, B. J. McCarthy, R. T. Schimpke, and A. Tissieres, 301–16. Life Sciences Research Report 4. Berlin: Dahlem Konferenzen.

Ehrenberg, C. G. 1838. *Die Infusionsthierchen als volkommene Organismen*. Leipzig.

Evans, Christopher. 1981. *The Micro-millennium*. New York: Washington Square Press.

Fahey, R. C.; Newton, G. L.; Arrick, B.; Overdale-Bogart, T.; and Aley, S. B. 1984. *Entamoeba histolytica:* A eukaryote without glutathione metabolism. *Science* 224:70–72.

Fleck, L. 1979. *Genesis and development of a scientific fact*. Chicago and London: University of Chicago Press.

Fracek, S. P., Jr. 1984. Tubulin-like proteins of *Spirochaeta bajacaliforniensis*, a new species from a microbial mat community at Laguna Figueroa, Baja California Norte, Mexico. Ph.D. diss., Boston University Graduate School.

Fracek, S. P., and Stolz, J. F. 1985. *Spirochaeta bajacaliforniensis* sp. n. from a microbial mat community of Laguna Figueroa, Baja California Norte, Mexico. *Archives for Microbiology* 142:317–25.

Friefelder, David. 1983. *Molecular biology: A comprehensive introduction to prokaryotes and eukaryotes*. Portola Valley, CA, and Boston: Jones and Bartlett.

Fulton, C. 1971. Centrioles. In *Origin and continuity of cell organelles*, ed. J. Reinert and H. Ursprung, 170–222. Heidelberg: Springer-Verlag.

Galleron, C. 1984. The fifth base: A natural feature of dinoflagellate DNA. *Origins of Life* 13:195–203.

Ghiselin, M. T. 1974. *The economy of nature and the evolution of sex*. Berkeley: University of California Press.

Giese, A. C. 1973. *Blepharisma:* The biology of a light-sensitive protozoan. Stanford: Stanford University Press.

Gillham, N. 1978. *Organellar heredity.* New York: Raven Press.

Girbardt, M., and Hadrich, H. 1975. Ultrastruktur des Pilzkernes III. Genese des Kernassoziierten Organells (NAO-"KCE"). *Zeitschrift für Allgemeine Mikrobiologie* 15:157–73.

Giusto, J. P., and Margulis, L. 1981. Karyotypic fission theory and the evolution of old world monkeys and apes. *BioSystems* 13:267–302.

Glaessner, M. F. 1984. *The dawn of animal life.* Cambridge: Cambridge University Press.

Goldknopf, I. L. 1978. Ubiquitin. In *The cell nucleus,* vol. 6, ed. H. Busch, 149–80. New York: Academic Press.

Goode, D. 1981. Microtubule turnover as a mechanism of mitosis and its possible evolution. *BioSystems* 14:271–87.

Goody, R. M., and Walker, J. C. G. 1972. *Atmospheres.* Englewood Cliffs, NJ: Prentice-Hall.

Gray, M. W. 1983. The bacterial ancestry of mitochondria. *BioScience* 33:693–99.

Gray, M. W., and Doolittle, W. F. 1982. Has the endosymbiotic theory been proven? *Microbiological Reviews* 46:1–42.

Grimes, G. W. 1976. Laser microbeam induction of incomplete doublets of *Oxytricha fallax. Genetical Research* 27:213–26.

———. 1980. Patterning and assembly of ciliature are independent processes in hypotrich ciliates. *Science* 209:281–83.

———. 1982. Nongenic inheritance: A determinant of cellular architecture. *BioScience* 32:279–80.

Grimes, G. W.; Knaupp-Waldrogel, E.; Goldsmith, C. M.; and Spoegler, M. 1981. Cytogeometrical determination of ciliary pattern formation in the hypotrich ciliate *Stylonychia mytilus. Developmental Biology* 84:477–80.

Halvorson, H. O., and Monroy, A., eds. 1985. *The origin and evolution of sex.* The Sixth Marine Biological Laboratory Symposium. New York: Alan R. Liss.

Hamilton, W. D. 1972. Altruism related phenomena, mainly in social insects. *Annual Review of Ecology and Systematics* 3:193–32.

Hapgood, F. 1979. *Why males exist: An inquiry into the evolution of sex.* New York: William Morrow.

Hartman, H. 1975. The centriole and the cell. *Journal of Theoretical Biology* 51:501–09.

Hartman, H.; Puma, J. P.; and Gurney, T. 1974. Evidence for the association of RNA with the ciliary basal bodies of *Tetrahymena. Journal of Cell Science* 16:241–60.

Haskins, E. F., and Therrien, C. D. 1978. The nuclear cycle of the myxomycete *Echinostelium minutum.* 1. Cytophotometric analysis of the nuclear DNA content of the amoebal and plasmodial phases. *Experimental Mycology* 2:32–40.

Haynes, R. H. 1985. Molecular mechanisms in genetic stability and change: The role of deoxyribonucleotide pool balance. In *Genetic consequences of nucleotide imbalance,* ed. Frederick J. de Serre, 1–23. New York: Plenum.

Heath, I. B. 1980a. Mechanisms of nuclear division in fungi. In *The fungal nucleus,* ed. K. Gull and S. Oliver, 85–112. Cambridge: Cambridge University Press.

———. 1980b. Variant mitosis in lower eukaryotes: Indication of the evolution of mitosis? *International Reviews of Cytology* 64:1–80.

Heidemann, S. R.; Sander, G.; and Kirschner, M. W. 1977. Evidence for a functional role of RNA in centrioles. *Cell* 10:337–50.

Herzog, M.; Von Boletsky, S.; and Soyer, M.-O. 1984. Ultrastructural and bio-chemical nuclear aspects of eukaryotic classification: Independent evolution of the dinoflagellates as a sister group of the actual eukaryotes? *Origins of Life* 13:205–15.

Hess, R. T., and Menzel, D. B. 1967. Rat kidney centrioles: Vitamin E intake and oxygen exposure. *Science* 159:985–87.

Hollande, A., and Caruette Valentin, J. 1972. Le problème du centrosome et la cryptopleuromitose atractophorienne chez *Lophomonas striata*. *Protistologica* 8:267–78.

Horgen, P. A., and Silver, J. C. 1978. Chromatin in eukaryotic microbes. *Annual Review of Microbiology* 32:249–84.

Hovind-Hougen, K. 1978. Determination by means of electron microscopy of mor-phological criteria of value for the classification of some spirochetes, in particu-lar treponemes. *Acta Pathologica et Microbiologica Scandinavica, Section B, Supplement*, no. 255.

Huang, B.; Ramanis, Z.; Dutcher, S. K.; and Luck, D. J. L. 1982. Uniflagellar mutants of *Chlamydomonas:* Evidence for the role of basal bodies in transmission of positional information. *Cell* 29:745–53.

Huxley, A. L. 1946. *Brave new world.* New York: Harper and Brothers.

Jackson, R. C., and Hauber, D. P., eds. 1983. *Polyploidy.* Benchmark Papers in Genetics. Stroudsburg, PA: Hutchinson Ross.

Jantsch, E. 1980. *The self-organizing universe: Scientific and human implications of the emerging paradigm of evolution.* Oxford and New York: Pergamon Press.

Jennings, H. S. 1920. *Life and death, heredity and evolution in unicellular organisms.* Boston: R. G. Badger.

Kaveski, S., and Margulis, L. 1983. The "sudden explosion" of animals about 600 million years ago: Why? *American Biology Teacher* 45:76–82.

Knoll, A., and Barghoorn, E. S. 1977. Archean microfossils showing cell division from the Swaziland System of South Africa. *Science* 198:396–98.

Knowles, C. J., ed. 1980. *The diversity of bacterial respiratory systems.* 2 vols. Boca Raton, FL: CRC Press.

Kornberg, A. 1980. *DNA synthesis.* San Francisco: W. H. Freeman.

Kuhn, Thomas. 1970. *The structure of scientific revolutions.* 2d ed. Chicago: Univer-sity of Chicago Press.

Kung, C.; Chang, S. Y.; Satow, Y.; Van Houten, J.; and Hansma, H. 1975. The genetic dissection of behavior in *Paramecium. Science* 188:898.

Langridge, J. 1982. Precambrian evolutionary genetics. In *Mineral deposits and the evolution of the biosphere,* ed. H. D. Holland and M. Schidlowski, 83–100. Berlin, Heidelberg, and New York: Springer-Verlag.

Larson, D. E., and Dingle, A. D. 1981. Development of the flagellar rootlet during *Naeglaria* flagellate differentiation. *Developmental Biology* 86:227–35.

Lauterborn, R. 1896. *Untersuchungen über Bau, Kernteilung und Bewegung der Diatomeen.* Leipzig: W. Engelman.

Law, R., and Lewis, D. H. 1983. Biotic environments and the maintenance of sex: Some evidence from mutualistic symbioses. *Biological Journal of the Linnean Society* 20:249–76.

Lederberg, J. 1955. Genetic recombination in bacteria. *Science* 122:920.

Lemischka, I., and Sharp, P. A. 1982. The sequences of an expressed rat α-tubulin gene and a pseudogene with an inserted repetitive element. *Nature* 300:330–35.

Lewin, R. 1984. No genome barriers to promiscuous DNA. *Science* 224:970–71.

Lidstrom, M. E.; Engebrecht, J.; and K. H. Nealson. 1983. Plasmid-mediated manganese oxidation by a marine bacterium. *Federation of European Micro-biological Societies Microbiology Letters* 19:1–6.

Little, M.; Ludueña, R. F.; Morejohn, C.; Asnes, C.; and Hoffman, E. 1984. The tubulins of animals, plants, fungi and protists: Implications for metazoan evolution. *Origins of Life* 13:169–76.

Lynn, D., and Small, E. B. 1981. Protist kinetids: Structural conservatism, kinetid structure and ancestral states. *BioSystems* 14:377–85.

———. n.d. Ciliophora. In *The Protoctista*, ed. L. Margulis, J. O. Corliss, and D. Chapman. Boston: Jones and Bartlett (forthcoming).

McCarty, M. 1985. *The transforming principle: Discovering that genes are made of DNA*. New York: W. W. Norton.

Margulis, L. 1974. On the evolutionary origin and possible mechanism of colchicine-sensitive mitotic movements. *BioSystems* 6:16–36.

———. 1975. Microtubules and evolution. In *Microtubules and microtubule inhibitors*, ed. M. Borgers and M. de Brabander, 3–18. Amsterdam: North-Holland.

———. 1981. *Symbiosis in cell evolution*. San Francisco: W. H. Freeman.

———. 1982. Microtubules in microorganisms and the origins of sex. In *Microtubules in microorganisms*, ed. P. Cappucinelli and R. Morris, 341–50. New York: Marcel Dekker.

Margulis, L.; Chase, D.; and Guerrero, R. 1986. Microbial communities. *BioScience* vol. 36, no. 3.

Margulis, L.; Chase, D.; and To, L. P. 1979. Possible evolutionary significance of spirochetes. *Transactions of the Royal Society of London, Series B* 204:189–98.

Margulis, L.; Corliss, J. O.; and Chapman, D., eds. n.d. *The Protoctista*. Boston: Jones and Bartlett (forthcoming).

Margulis, L.; Lopez-Baluja, L.; Awramik, S. W.; and Sagan, D. 1986. Community living long before man: Fossil and living microbial mats and early life. In *Man's effect on the global environment*, ed. A. A. Orio and D. Botkin. Amsterdam: Elsevier.

Margulis, L., and McKhann, H. n.d. Ultrastructure of *Paratetramitus* sp. from a Cuban microbial mat. In preparation.

Margulis, L.; Mehos, D.; and Kaveski, S. 1983. There is no such thing as a one-celled animal or plant. *Science Teacher* 50:34–36, 41–43.

Margulis, L., and Obar, R. 1985. *Heliobacterium* and the origin of chrysoplasts. *BioSystems* 17:361–65.

Margulis, L., and Sagan, D. 1986. *Microcosmos: Four billion years of evolution from our microbial ancestors*. New York: Summit.

Margulis, L., and Schwartz, K. V. 1982. *Five kingdoms*. San Francisco: W. H. Freeman.

Margulis, L.; Soyer-Gobillard, M.-O.; and Corliss, J., eds. 1984. *Evolutionary protistology: The organism as cell*. Dordrecht and Boston: D. Reidel.

Margulis, L., and Stolz, J. F. 1985. Cell symbiosis theory: Status and implications for the fossil record. *Advances in Space Research. COSPAR Proceedings* 4:195–201.

Margulis, L.; To, L. P.; and Chase, D. 1981a. The genera *Pillotina, Hollandina* and *Diplocalyx*. In *The prokaryotes*, ed. M. P. Starr et al., 1:548–54. Berlin and New York: Springer-Verlag.

———. 1981b. Microtubules, undulipodia and pillotina spirochetes. *Annals of the New York Academy of Sciences* 361:356–68.

Margulis, L.; Walker, J. C. G.; and Rambler, M. B. 1976. A reassessment of the roles of oxygen and ultraviolet light in Precambrian evolution. *Nature* 264:620–24.

Marrs, B. L. 1983. Genetics and molecular biology. In *The phototrophic bacteria:*

*Anaerobic life in the light,* ed. J. G. Ormerod, 186–214. Oxford: Blackwell Scientific Publishers.

Martin, Jesus. 1984. Cortical experiments with hypotrichous ciliates. Personal communication.

Maynard Smith, J. 1978. *The evolution of sex.* Cambridge: Cambridge University Press.

Mereschkovsky, C. 1905. Le plante considerée comme une complex symbiotique. *Bulletin Société Science Naturelle, Ouest* 6:17–98.

Miller, S. L., and Orgel, L. E. 1974. *Origins of life on Earth.* Englewood Cliffs, NJ: Prentice-Hall.

Mills, D. 1984. Q-beta replicase. Personal communication.

Mizukami, I., and Gall, J. 1966. Centriole replication. II. Sperm formation in the fern, *Marsilea,* and the cycad, *Zamia. Journal of Cell Biology* 29:97–111.

Moestrup, O. 1982. Flagellar structure in algae. *Phycologia* 21:427–528.

Monaud, M., and Pappas, G. D. 1968. Cilia formation in the adult cat brain after pargyline treatment. *Journal of Cell Biology* 35:599–602.

Muller, H. J. 1930. Radiation and genetics. *American Naturalist* 64:220–25.

Muscatine, L., and Neckelmann, N. 1983. Regulatory mechanisms maintaining the *Hydra-Chlorella* symbiosis. *Proceedings of the Royal Society of London, Series B* 219:193–210.

Nomura, M. 1984. The control of ribosome synthesis. *Scientific American* 250:102–14.

Obar, R. 1985. Purification of tubulin-like proteins from a spirochete. Ph.D. diss., Boston University Graduate School.

Obar, R., and Green, J. 1985. The molecular evolution of mitochondria. *Journal of Molecular Evolution* (submitted).

Omodeo, P. 1975. Evolution of the genome considered in the light of information theory. *Bollettino di Zoologia, Agraria e di Bachicoltura* 42:351–79.

Omodeo, P.; Capanna, E.; and Pallini, V. 1982. Cell and contractile protein evolution. In *Cell growth,* ed. Claudio Nicolini. New York: Plenum.

Page, F. 1983. *The marine gymnoamoebae.* Cambridge: Institute of Terrestrial Ecology.

Pickett-Heaps, J. D. 1971. The autonomy of the centriole: Fact or fallacy? *Cytobios* 3:205–214.

———. 1974. Evolution of mitosis and the eukaryote condition. *BioSystems* 6:37–48.

Pickett-Heaps, J. D.; Tippett, D. H.; and Andreozzi, J. A. 1978. Cell division in the pennate diatom *Pinnularia.* V. Observations on live cells. *Biologie Cellulaire* 35:295–304.

Pontecorvo, G. 1958. *Trends in genetic analysis.* New York: Columbia University Press.

Porter, David L. n.d. Labyrinthulamycota. In *Handbook of protoctista,* ed. L. Margulis, J. O. Corliss, and D. Chapman. Boston: Jones and Bartlett (forthcoming).

Prescott, D.; Marti, R. G.; and Bostock, C. J. 1973. Genetic apparatus of *Stylonychia sp. Nature* 242:576, 597–600.

Pyrozinski, K., and Malloch, D. 1975. On the origin of land plants: A question of mycotrophy. *BioSystems* 6:153–64.

Radding, C. M. 1978. Genetic recombination: Strand transfer and mismatch repair. *Annual Review of Biochemistry* 47:347–88.

Raikov, I. B. 1969. The macronucleus of ciliates. *Research in Protozoology* 3:1–128.

———. 1972. Nuclear phenomena during conjugation and autogamy in ciliates. *Research in Protozoology* 4:147–289.

———. 1982. *The protozoan nucleus.* Heidelberg and New York: Springer-Verlag.

Rambler, M. B. 1980. Ultraviolet light and oxygen in the pre-Phanerozoic. Ph.D. diss., Boston University Graduate School.

Rambler, M. B., and Margulis, L. M. 1980. Bacterial resistance to UV radiation under anaerobiosis: Implications for pre-Phanerozoic evolution. *Science* 210:638–40.

Randall, Sir J., and Disbrey, C. 1965. Evidence for the presence of DNA at basal body sites in *Tetrahymena pyriformis. Proceedings of the Royal Society of London, Series B* 162:473–91.

Randall, Sir J.; Warr, J. R.; Hopkins, J. M.; and McVittie, A. 1964. A single gene mutation of *Chlamydomonas reinhardi* affecting motility: A genetic and electron microscope study. *Nature* 204:912–14.

Richmond, M. H., and Smith, D. C. 1979. *The cell as a habitat.* London: The Royal Society.

Ris, H. 1961. Ultrastructure and molecular organization of genetic systems. *Canadian Journal of Genetics and Cytology* 3:95–120.

————. 1975. Primitive mitotic systems. *BioSystems* 7:298–304.

Roberts, J. W.; Roberts, C. W.; Craig, N. L.; and Phizicky, E. M. 1985. Activity of the *Escherichia coli* recA-gene product. *Biochemistry* 24:917–20.

Roos, U.-P. 1984. From promitosis to mitosis: An alternative hypothesis on the origin and evolution of the mitotic spindle. *Origins of Life* 13:183–93.

Rose, M. R. 1983. The contagion mechanism for the origin of sex. *Journal of Theoretical Biology* 101:137–46.

Rosson, R., and Nealson, K. H. 1982a. Manganese bacteria and the marine manganese cycle. In *The environment of the deep sea,* ed. J. G. Morin and W. G. Ernst, 201–16. Englewood Cliffs, NJ: Prentice-Hall.

————. 1982b. Manganese binding and oxidation by spores of a marine bacillus. *Journal of Bacteriology* 151:1037–44.

Sagan, D., ed. 1985. *The global sulfur cycle: Planetary biology and microbial ecology.* NASA Technical Memorandum 87570.

Saier, M., and Jacobson, G. 1984. *The molecular basis of sex and differentiation: A comparative study of evolution, mechanism and control in microorganisms.* Heidelberg and New York: Springer-Verlag.

Savage, J. R. K.; Cawood, A. H.; and Rapworth, D. G. 1983. The disparity between homologous chromosomes during DNA replication. *Journal of Theoretical Biology* 100:631–43.

Schenk, H. E. A., and Schwemmler, W., eds. 1983. *Endocytobiology: Intracellular space as oligogenetic ecosystem.* Vol. 2. Berlin: Walter de Gruyter.

Schiff, J. A. 1981. Origin and evolution of the plastid and its function. *Annals of the New York Academy of Sciences* 361:166–92.

Schuster, F. L. 1963. An electron microscope study of the amoeboflagellate *Naeglaria gruberi.* I. The amoeboid and flagellate stages. *Journal of Protozoology* 10:297–312.

Schuster, P., and Sigmund, K. 1982. A note on the evolution of sexual dimorphism. *Journal of Theoretical Biology* 94:107–10.

Schwemmler, W. 1980. Endocytobiology: A modern field between symbiosis and cell research. In *Endocytobiology,* ed. W. Schwemmler and H. E. A. Schenk, 1:943–67. Berlin: Walter de Gruyter.

Searcy, D. 1982. *Thermoplasma:* A primitive cell from a refuse pile. *Trends in Biochemical Science (TIBS)* 7:183–85.

Searcy, D. G., and Stein, D. B. 1980. Nucleoplasm subunit structure in an unusual prokaryotic organism: *Thermoplasma acidophilum. Biochimica et Biophysica Acta* 609:108–95.

Searcy, D. G.; Stein, D. B.; and Green, G. R. 1978. Phylogenetic affinities between eukaryotic cells and a thermoplasmic mycoplasma. *BioSystems* 10:19–28.

Silver, J. 1984. Histone proteins and chromatin of fungi [personal communication]. Department of Biology, Scarborough College, University of Toronto.

Silver, S. 1983. Bacterial interactions with mineral cations and anions: Good ions and bad. in *Biomineralization and biological metal accumulation,* ed. P. Westbroek and E. W. de Jong, 439–57. Dordrecht: D. Reidel.

Smith-Sonneborn, J., and Plaut, W. S. 1969. Studies on the autonomy of pellicular DNA in *Paramecium. Journal of Cell Science* 5:365–72.

Sonea, S., and Panisset, P. 1983. *The new bacteriology.* Boston: Jones and Bartlett.

Sorokin, S. P. 1968. Reconstruction of centriole formation and ciliogenesis in mammalian lungs. *Journal of Cell Science* 3:207–36.

Soyer-Gobillard, M.-O. 1982. *Titres et travaux scientifique.* Laboratoire Arago, Centre National de Recherche Scientifique, 15 Quai Anatole France, Paris.

Spiegel, F. W. 1981. Phylogenetic significance of the flagellar apparatus in protostelids (Eumycetozoa). *BioSystems* 14:491–99.

Stack, S. M., and Brown, M. V. 1969. Somatic pairing, reduction and recombination: An evolutionary hypothesis of meiosis. *Nature* 222:1275–76.

Starr, M. P.; Stolp, H.; Trueper, H. G.; Balow, A.; and Schlegel, H. G., eds. 1983. *The prokaryotes: A handbook on habitats, isolation and identification of bacteria.* 2 vols. Heidelberg and New York: Springer-Verlag.

Stern, H., and Hotta, Y. 1967. Chromosome behavior during development of meiotic tissue. In *The control of nuclear activity,* ed. L. Goldstein, 47–76. Englewood Cliffs, NJ: Prentice-Hall.

Stolz, J. F. 1984. The effects of catastrophic inundation on a microbial mat at Laguna Figueroa, Baja California Norte, Mexico. Ph.D. diss., Boston University Graduate School.

Strogatz, S. 1983. Topology of zigzag chromatin. *Journal of Theoretical Biology* 103:601–07.

Stryer, L. 1981. *Biochemistry.* San Francisco: W. H. Freeman.

Stubblefield, E., and Brinkley, B. R. 1966. Cilia formation in Chinese hamster fibroblasts *in vitro* as a response to Colcemid treatment. *Journal of Cell Biology* 30:645–52.

Sulloway, F. J. 1983. *Freud: Biologist of the mind.* New York: Basic Books.

Suzuki, D.; Griffiths, A. J. F.; and Lewontin, R. C. 1981. *Introduction to genetic analysis.* 2d ed. San Francisco: W. H. Freeman.

Tartar, V. 1961. *The biology of* Stentor. Oxford and New York: Pergamon Press.

Taylor, F. J. R. 1980. On dinoflagellate evolution. *BioSystems* 13:65–108.

———. 1983. Some ecoevolutionary aspects of symbiosis. *International Review of Cytology, Supplement* 14:1–28.

Thorington, G. 1978. The symbionts of *Hydra viridis.* Ph.D. diss., Boston University Graduate School.

Thornley, A. L., and Harington, A. 1981. Diploidy and sex as the selective advantage for retaining genes transferred from mitochondria and plastid ancestors in the nuclear genome. *Journal of Theoretical Biology* 91:515–23.

To, L. P. 1978. The large pillotina spirochetes of the dry wood termites and their microtubules. Ph.D. diss., Boston University Graduate School.

To, L. P.; Margulis, L.; Chase, D.; and Nutting, W. L. 1980. The symbiotic microbial community of the Sonora desert termite: *Pterotermes occidentis. BioSystems* 13:109–37.

Todd, N. B. 1970. Karyotypic fissioning and carnivore evolution. *Journal of Theoretical Biology* 26:445–80.

Turner, F. R. 1968. An ultrastructural study of plant spermatogenesis. Spermatogenesis in *Nitella*. *Journal of Cell Biology* 37:370–93.

Van Bruggen, J. J. A.; Stumm, C. K.; and Vogels, G. D. 1983. Symbiosis of methanogenic bacteria and sapropelic protozoa. *Archives of Microbiology* 136:89–95.

Varela, F., and Maturana, H. R. 1974. Autopoiesis: The organization of living systems, its characterization and a model. *BioSystems* 5:187–96.

Vidal, G. 1984. The oldest eukaryotic cells. *Scientific American* 250:48–57.

Vidal, G., and Knoll, A. H. 1983. Proterozoic plankton. *Geological Society of America, Memoir* 161:265–77.

Waddington, C. 1976. Concluding remarks. In *Evolution and consciousness*, ed. E. Jantsch and C. H. Waddington, 243–49. Reading, MA: Addison Wesley.

Walker, J. C. G. 1986. *History of the Earth's atmosphere*. Boston: Jones and Bartlett.

Wallin, I. 1927. *Symbionticism and the origin of species*. Baltimore: Williams and Wilkins.

Watanabe, T. 1977. Chemical properties of mating substances in *Paramecium caudatum*: Effect of various agents on mating reactivity of detached cilia. *Cell Structure and Function* 2:241–47.

Watson, A. J., and Lovelock, J. E. 1983. Biological homeostasis of the global environment: The parable of Daisyworld. *Tellus* 35B:284–89.

Weinberg, S. 1977. *The first three minutes*. New York: Basic Books.

Whatley, J., and Chapman-Andresen, C. n.d. *Pelomyxa*. In *The Protoctista*, ed. L. Margulis, J. O. Corliss, and D. Chapman. Boston: Jones and Bartlett (forthcoming).

Wheatley, D. N. 1982. *The centriole: A central enigma of cell biology*. Amsterdam, NY, and Oxford: Elsevier Biomedical Press.

Williams, G. C. 1966. *Adaptation and natural selection*. Princeton: Princeton University Press.

Wilson, E. B. 1925. *The cell in heredity and development*. New York: Macmillan.

Wilson, E. O. 1975. *Sociobiology: The new synthesis*. Cambridge: Harvard University Press.

———. 1978. *On human nature*. Cambridge: Harvard University Press.

Witkin, E. M. 1969. Ultraviolet-light induced mutation and DNA repair. *Annual Review of Microbiology* 23:487–514.

Witzmann, R. F. 1981. *Steroids: Keys to life*. New York: Van Nostrand Reinhold.

Woese, C. R. 1981. Archaebacteria. *Scientific American* 244:98–122.

Woodcock, C. L. F., and Bogorad, L. 1970. Evidence for variation in the quantity of DNA per plastid in *Acetabularia*. *Journal of Cell Biology* 44:361–75.

Wright, R. M., and Cummings, D. J. 1983. Integration of mitochondrial gene sequences within the nuclear genome during senescence in a fungus. *Nature* 302:86–88.

Younger, K. B.; Banerjee, S.; Kelleher, J. K.; Winston, M.; and Margulis, L. 1972. Evidence that the synchronized production of new basal bodies is not associated with DNA synthesis in *Stentor coeruleus*. *Journal of Cell Science* 11:621–37.

Zahn, R. K. 1984. A green alga with minimal eukaryotic features: *Nanochlorum eukaryoticum*. *Origins of Life* 13:289–303.

# INDEX

*Italicized entries refer to illustrations.*